Emanuely Velozo Aragão Bueno
Dante Alves Medeiros
Marcelo Luiz Chicati

Analysing urban sprawl using a spatial separation indicator

Emanuely Velozo Aragão Bueno
Dante Alves Medeiros
Marcelo Luiz Chicati

Analysing urban sprawl using a spatial separation indicator

Urban expansion and the use of a spatial separation indicator

ScienciaScripts

Cover image: www.ingimage.com

This book is a translation from the original published under ISBN 978-613-0-17275-6.

Publisher:
Sciencia Scripts
is a trademark of
Dodo Books Indian Ocean Ltd. and OmniScriptum S.R.L publishing group

120 High Road, East Finchley, London, N2 9ED, United Kingdom
Str. Armeneasca 28/1, office 1, Chisinau MD-2012, Republic of Moldova, Europe
Managing Directors: Ieva Konstantinova, Victoria Ursu
info@omniscriptum.com

Printed at: see last page
ISBN: 978-620-8-60161-4

ACKNOWLEDGEMENTS

Firstly to God for all the graces I have received and for guiding me at every moment of my life.

To my parents Iraci de Souza Velozo and Sergio Castro Aragão and my siblings Franciely Velozo Aragão, Renan Velozo Aragão and Maria Rita Velozo Aragão for all their care and love.

To my grandfather Durvalino Rafael Velozo (In Memorian) for his support, affection and learning.

To my husband Reginaldo de Souza Bueno for his patience and support in overcoming the stages of my personal and professional growth.

To Professors Dante Alves Medeiros Filho and Marcelo Chicati for their learning and patience in carrying out this research.

To my fellow master's students for having shared so many professional experiences and for the new friendships made.

To all the professors of the Master's Programme in Urban Engineering - PEU and to Douglas for the opportunity to study for the master's degree.

Capes and Fundação Araucária for the financial incentive to carry out the research.

Finally, to everyone who helped me carry out my research and complete my master's degree.

"Willpower has no set day, no right moment, much less does it need a miracle exercise. Willpower is your ability to have focus, persistence, discipline and love for your journey. "

Zora Viana

SUMMARY

Cities are currently suffering from the effects of urban growth and these reflexes can be clearly seen in the way some populations live and in the spatial dynamics of each region. Brazilian cities are unprepared for this growth, unlike other countries that are investing heavily in measures to minimise these urban impacts. Lack of access, conurbation, places with inadequate infrastructure and urban voids are the main problems encountered in urban areas. Given these problems, the main objective of this research was to assess the applicability of the Allen index in identifying the shortest access distances to the central region of Campo Mourão, PR in urban expansion studies. This study used a **spatial separation** indicator **called the "AUen index", which** verifies the effects of urban sprawl on the access of users from different locations in the urban network to the central region. The results of the index are found using a mathematical formula that allows the best levels of access to be calculated for different points, which shows which years or decades have seen the greatest growth in streets and the best access. As allotments grew, intersections increased and distances of between 4.5 km and 7.5 km were used, while in the neighbourhoods established between 1950 and 1970 the distances between 1.5 km and 4.5 km gradually decreased. The 1960s saw the greatest peak in growth, with 1009 intersections. After this period, there was a gradual decrease and it was only in the 2010s that these figures began to increase, with 11 intersections, a figure higher than in the 2000s. The results therefore validate the use of the spatial separation indicator for a medium-sized city.

Keyword: Spatial separation, Access, Indicator.

SUMMARY

CHAPTER 1

INTRODUCTION

The urban space is constantly changing and this is reflected in the quality of life of the inhabitants and in the decentralisation caused by the demand for buildings and the concentration of populations in a given region. According to data from the IBGE (Brazilian Institute of Geography and Statistics) 2010 census, Brazil's urbanisation rate averaged 84.3%, while the state of Paraná showed 85.3%, which shows that the state is booming.

Urbanisation rates reflect the attractions found in the state, such as job opportunities, quality of life and infrastructure, all of which are relevant to expanding populations in its cities.

The aspects involved in the proper development of urban networks must be included in public policies and in the culture of the residents of each region. Access is one of them, and must be studied to minimise the unfavourable impacts of accelerated urbanisation. The restructuring of urban networks is linked to urban mobility in municipalities and major capitals.

The growth of cities can be analysed using indicators that make it possible to identify the possible problems hindering their development. The use of appropriate indicators makes it easier to understand problematic situations in the urban environment that would not be directly perceived by their managers without the appropriate instruments.

In order to work on the favourable characteristics of cities, prior studies must be carried out by various professionals. Preliminary knowledge of needs combats flaws that could be detrimental to the city as a whole.

The need to accommodate the ever-growing population in large centres attracts the setting up of allotments, which are inserted into available sites within the urban fabric. Generally, these allotments are located far from the city centre. In the last fifteen years, the city of Campo Mourão, where the research was carried out, has seen 27 allotments set up. Thus, for each year, two new allotments were approved by the city council.

In this sense, various factors should be reviewed and prioritised, taking into account the local characteristics of each region. It is therefore important to propose a study focused on urban planning that identifies the needs of society, as well as presenting data that can be used by managers for further research.

1.2 OBJECTIVE

The general objective of this work is to assess the applicability of the Allen index in identifying the shortest access distances to the central region of Campo Mourão, PR in urban expansion studies.

1.2.1 SPECIFIC OBJECTIVES

- Drawing up maps of the road system, urban sprawl, intersections and the Allen index;

- Identifying the decades and years with the smallest distances between access by calculating the Allen index;
- Demonstrate whether Allen's index was compatible for the city analysed.

1.3 BACKGROUND

Cities are experiencing a boom in population. However, urban planning problems are recurrent in cities with constant urban expansion. Some locations are planned when they are established, but the lack of periodic studies interferes with adequate infrastructure for new constructions.

New housing developments are becoming increasingly remote from the town centres. This makes it difficult for inhabitants to access the basic requirements for their survival, which is reflected in the lack of schools, crèches and basic health units for each existing region.

Allotment companies, together with municipal bodies, should present a project not only for the plots to be built on, but also for basic resources for the future population. In this way, the problem of access to service points can be reduced.

All these factors caused by disorganised urban sprawl interfere not only with the structure of cities, but also with the lives of the population. These users end up needing more time to travel to certain places, such as schools and other public institutions.

Urbanisation has brought countless benefits to its residents and to the development of the city itself. However, rapid growth has brought with it various consequences, such as poor quality of life and a collapse in the lives of the inhabitants (BARRIQUELLO, 2011). This totally interferes with the healthy growth cycle of the municipality and affects its economy.

These disorganised changes that cities undergo cannot be predicted, due to a number of conditions. Practically all cities show transformations and wear and tear on their natural environments, because man's practice of conducting and sustaining his needs favours these changes (BARIQUELLO, 2011).

This makes it essential to understand the development of municipalities and analyse possible solutions to alleviate these growing issues. Knowing and observing the development of the municipality under analysis since its foundation, it is possible to demonstrate its evolution.

In this way, research into urban sprawl that deals with knowledge of the development of cities serves to understand what factors have led to this situation, serving as a basis for municipal bodies to put into practice possible solutions that have been questioned. The study sought to deal with these factors in a clear and objective way, so that it can serve as support material for municipal bodies in resolving access problems. In addition, the tool used is simple to apply and its results are objective. These are important factors, as the cost-effectiveness of using a low-cost mechanism is essential.

In order to find out about these factors, the research presents a study using data obtained from the cartographic base of the city of Campo Mourão - PR, made available by the City Hall, where they were georeferenced using the ARCGIS tool. The results were analysed using the **spatial separation indicator** "Allen's Index", which made it possible to verify the locations in the city with the best access and allowed them to be compared.

CHAPTER 2

LITERATURE REVIEW

2.1 URBANISATION

Brazilian cities are constantly changing and these changes contribute to disorganised growth in urban areas. Urbanised areas show changes in their urban fabric due to the fact that expansion is the result of the population's needs and the space that is available. Oliveira (2006) reports that urban development is linked to the type of economic activity that predominates in certain regions.

Large educational centres also develop municipalities, as they become benchmarks in various areas of activity. As a result, population growth increases, mainly influencing the municipality's economy. To this end, the endogenous theory emphasises the important role of municipalities in diagnosing the exact time to reformulate strategies to increase and control their growth rates. It becomes clear which sector has the necessary tools for this change (COSTA, et al. 2013).

The factors influencing this diagnosis are the discovery of spending on health, education, culture and sanitation in the municipalities. Knowledge of investment in these areas contributes to and affects the expansion of cities. The departments that carry out this work must have trained staff who can supervise and carry out their duties correctly.

According to Costa et al. (2013), the theory of endogenous growth believes in the power of government influence over the economy, thanks to the numerous studies carried out in this area. Other theories reinforce this idea, such as Krugman's theory known as the new economic geography (NGE).

The main objective of NGE is to emphasise that the growth of a certain place is linked to the existing activities of capitals, states, regions and so on. The theory believes in the existing forces that cause economic activities to be allocated according to the forces formed by the existing characteristics of each space. These forces can be known as centrifugal, which are responsible for the dispersion of economic activities and centripetal forces known for concentration (COSTA, et al. 2013).

Industries are considered a centripetal force, as they attract other companies to their vicinity, making the area more industrialised with various agglomerations. Another example is found in transport costs, because when these companies are set up, the concern to reduce costs makes them set up nearby in order to minimise the cost of obtaining raw materials (COSTA et al., 2013).

The centrifugal forces known to cause cities to disperse can be triggered by a number of factors, including lack of employment, rampant crime, lower wages and other well-known factors. All cities have these characteristics, but some municipalities do not have adequate infrastructure for their residents and this disperses their population to other locations (COELHO, 2013).

Centrifugal forces will always be present, whether in large cities or small towns. In order to minimise these forces, the departments responsible for urban planning must be present at so that they can monitor the new allotments to be built, as well as the predefined locations for new homes with adequate resources (COSTA et al., 2013).

The country, cities and states will be effectively developed if the actions and measures to be taken are properly planned. When the objectives to be achieved are determined, the search to solve or minimise them becomes more effective (RIBEIRO, 2011).

Ribeiro (2011) also mentions that cities are formed by the population and change according to the needs of their inhabitants. That's why their spatial organisation is constantly changing.

The City Statute prioritises the main urban characteristics, promoting their main sources of economy. In this way, valorisation contributes to its continuous evolution (RIBEIRO, 2011). The main objective of Law No. 10.257 of 10 July 2001 is to establish guidelines to guide urban growth and land use and occupation in order to create a fairer and more sustainable society. Article 182 of the City Statute emphasises this in the first paragraph:

> The masterplan, approved by the city council and compulsory for cities with more than 20,000 inhabitants, is the basic instrument of development and expansion policy.

The state of Paraná underwent major changes in its urbanisation in the second half of the 20th century. This had negative consequences for the environment, people and government (MOURA, 2004). Brazil, unlike other countries, has suffered abruptly from urbanisation and its most affected areas have been demographic, social and ecological (FARIA, 1991 apud MOURA, 2004).

According to Moura (2004), in 1950 the state of Paraná had an estimated population of 2.1 million inhabitants and this growth more than doubled in just two decades. It grew to 7 million inhabitants in 1970. Its constant evolution meant that between 1991 and 2000 it reached approximately 9.5 million people (MOURA, 2004). According to the Paraná Institute for Economic and Social Development (2014), the state has an estimated population of 11,081,692, with an urbanisation rate of 85.33%. According to the 2010 census, the rural population is 1,531,834.

All people who come from rural areas or small towns end up taking their customs with them to the places where they settle. When they join the urban environment, they tend to add to its urban values in all possible areas, especially in the competition for jobs. The main consequences left behind in the urban area are the aforementioned changes to the urban fabric and the increase in population density in existing neighbourhoods (MOURA, 2004).

2.1.1 URBAN SPRAWL

According to Araújo et al. (2013), urban areas have experienced unbridled growth in recent decades, which can be seen in the countless settlements that lack adequate infrastructure for their residents. These clandestine installations not only interfere with the growth of cities but also with the environment.

Lima (1998) reports that the expansion of Brazilian cities follows a grid format, so that their layout is regular. This is how it develops as people move in. Generally, the main point of expansion is marked by the existing parish church or central squares. Thus the main streets and avenues are concentrated in the area considered to be the centre, as this is where the main events take place.

Silva (2014) reports that the accelerated and disordered growth is a result of the Brazilian economy, due to the mechanisation imposed on agriculture. This was due to the high consumption of products and the time spent on production, as well as the expansion of industries in the 1970s.

POLIDORO (2012) points out that urban sprawl is taking new directions, where rural and urban areas are completely united, making it difficult to recognise the limits of occupation in some regions, and it is necessary to know and discover new effective and judicious mechanisms that minimise these phenomena of disordered expansion.

Polidoro (2011) also states that urban sprawl slows down the growth of cities and induces the emergence of conurbations and physical agglomeration of municipalities, so that regulatory laws do not apply only to one region but to the whole formed by growth. This would increase conflicts of jurisdiction, as federal, state and municipal roads meet, causing problems for the municipality to resolve.

2.1.2 URBAN SPRAWL IN CAMPO MOURÃO

According to PROCÓPIO (2007 apud CRUZ, 2010) the migration process to the Campo Mourão - PR region began in the 1940s. The major factor driving this search was the availability of cheap land, low-interest financing and the existence of an agricultural cooperative, as the region is predominantly used to growing soya and its climate contributes to this type of agriculture.

Figure 01 is an aerial photo from 1953, made available by the Planning Department, showing the city's first blocks. At the start of urbanisation it was approximately 1.4 km^2, which today represents an extension of the city's streets in a northwest/southeast direction between Goioerê and Irmãos Pereira avenues and in the opposite southeast/northwest direction between Panambi and Edmundo Mercer streets (MARCOTTI, A.R.; MARCOTTI, T.C.B, 2011).

Figure 1: 1953 aerial photograph of the city of Campo Mourão - PR
Source: Municipality of Campo Mourão - PR.

The municipality's urban fabric is laid out on a plateau and is divided by a main spur that runs in a north-west/south-east direction. This means that they are separated by two drainage sub-basins, one of which collects water for the inhabitants of Campo Mourão - PR (CRUZ, 2010).

When Campo Mourão was first urbanised, the blocks were laid out in larger sizes due to the availability of land. Initially, the blocks were 118 metres by 60 metres, with a total of 18 plots per block (MARCOTTI, A.R.; MARCOTTI, T.C.B, 2011).

Population growth over the years has undergone major changes due to various factors already mentioned. In the case of the city of Campo Mourão - PR there has been a significant increase in the degree of urbanisation over the years, comparing the years 1940 to 2010 (Table 01). A number of changes have occurred with the increase in the number of inhabitants, with the rural population falling by 27,594 thousand inhabitants, a reflection of migration and the high technology that agriculture has become.

Table 1: Population Data and Degree of Urbanisation in the Municipality of Campo Mourão - PR

Year	Total Population	Rural Population	Urban Population	Degree of Urbanisation (%)
1940	11.964	-	-	-
1950	33.949	32.112	836	2,46%
1960	140.362	120.873	19.489	13,88%
1970	77.118	49.207	27.911	36,19%
1980	75.423	26.084	49.339	65,41%
1990	82.318	9,983	72.335	87,87%
2000	80.476	5.722	74.754	92,88%
2010	87.194	4.518	82.676	94,81%

Source: IBGE - CENSO (2010); VEIGA (1999)
Organised: MORIGI, Josimari de Brito (2013).

The urbanisation rate of some cities can either increase in line with their economic activity or decrease in the face of the problems they face. It is calculated through the ratio of the number of people living in the urban area to the total territorial population. The percentage of immigrants in Campo Mourão is estimated at over 8,800 people (MARIA, et al., 2011).

Campo Mourão's Municipal Human Development Index (MHDI) is high when compared to the Brazilian MHDI (Figure 02). This index aims to measure the degree of economic development and the quality of life offered to the population. It also reports on the development of cities according to the UNDP (United Nations Development Programme).

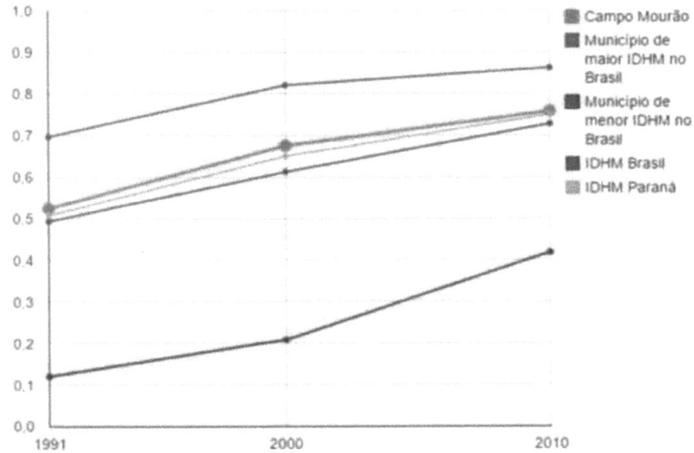

Figure 2: Evolution of the HDI
Source: UNDP, IPEA and FJP (2010).

2.2 URBAN SPACE AND ITS CHALLENGES

According to Correa (1989), urban space is characterised by different land uses arranged in relation to each other, being used for different locations. Whether in city centres, places involving commercial activities, residential and industrial areas, among others. All these elements that make up the set of land uses are known as the spatial organisation of the city.

For Correa (2009), there are agents who produce the space of a city, building or complementing the city according to their needs and priorities:

- Owners of the production area (large industries): they are major consumers of space, as they need larger premises at an affordable price. They also need to be located in places that are easily accessible to the population and close to railway lines to transport their products. There is also a relationship between the owners of production land and urban land. This is because industrial plots are cheap and large, unlike urban areas where plots are smaller and prices change abusively due to real estate speculation.

- Landowners: their main objective is to acquire greater land income, which they can benefit from by investing in companies or public organisations. Their interest is not in the use of the land but in the value that can be exchanged. Generally, these entrepreneurs sell to large developments or create high-cost allotments with a focus on gated communities.
- Property owners: they can carry out financing, development, property construction, marketing or technical studies. Developers who carry out this type of work intensify spatial segregation between populations. This makes it clear that there is room for more sophisticated properties, unlike the lower-income population, because their work becomes unequal.
- State: the main objective is to organise cities spatially, presenting instruments that can contribute to their development. Investing in the production of urban space, regulating land use, among other endeavours.
- Excluded social groups: these are people who are unable to pay rent for adequate housing, or even buy their own property due to a lack of jobs and other life opportunities. In this way, this group becomes excluded from the other classes, even though they use a large part of the urban space.

The division of spaces is imposed by the residents themselves. This fragmentation differentiates classes, valorises some areas and at the same time devalues others. This is caused by the intention to bring certain people or objects closer or further away. In this way, the poor are held back by the powerful, increasing their accumulation of capital and the processes of exploitation (CUNHA, 2011).

2.2.1 DILEMMAS OF CITIES: URBAN VOIDS AND DISPERSION

One of the dilemmas faced by many cities is known as spatial segregation, where there is a notorious division of classes in municipalities, states and all regions. These divisions can be social, political and religious (CORREIA, 1989).

Negri (2011) emphasises this idea of separation, as this problem has been known for a long time, since antiquity, when peoples were separated by their peculiar characteristics. Marcuse (2004 apud NEGRI, 2011) describes these divisions in stages:

- Cultural division: distinguished by language, local culture, ethnic characteristics, country, region or nationality.
- Functional division: is a division according to the activity carried out, being by rural and urban areas, residential and industrial.
- Division by hierarchical differences: this is the division of the power relationship that exists in cities.

Urban spaces, despite their increasing development, have urban voids due to various factors, but mainly due to property speculation, caused by the interest in increasing the value of land for a higher profit. According to Jordão (2013), cities are gradually adapting to capitalist standards, where one of the main concepts centres on consuming more and more goods and products.

According to Oliveira et al. (1991) the capitalist increase in land use is a consequence of the process of defining the price of urban land. This is one of the main factors responsible for reformulating property prices and changing spatial configurations.

These urban voids can be found at various points in an urban network, from the outskirts to the city centres. In the central region, as mentioned above, they are caused by an interest in increasing the value of property. In the case of more remote regions, the main aggravating factor is violence and distant routes for day-to-day activities.

However, as plots in central regions are expensive, the opportunity to acquire their own property is a fundamental need for any family group. Some people invest in new housing developments, which are located in more distant areas, but which are more accessible to their financial circumstances and have basic infrastructure for their operation.

Capitalist activities consider private land holdings to be a barrier to growth. They are seen as an investment opportunity for both cities and the property sector. However, they cannot be managed because they are private properties (OLIVEIRA et al., 1991).

According to Raia (1995 apud LIMA, 1998), urban voids are formed when new allotments are set up far from existing ones. And these spaces between them form urban voids that tend to increase in value due to the benefits that exist around them.

These new allotments form what is known as dispersion in the urban fabric. Companies expand according to the availability of land, and generally these new developments are located far from the centre. (LIMA 1998) emphasises that the increase in distance contributes to an increase in the cost of transport and the time spent travelling.

It takes longer for people to get from their homes to their jobs and educational institutions. Municipal bodies must therefore be prepared to provide the necessary support to their population, so that there are fewer conflicts at municipal, regional and state level.

According to Silva (1993 apud LIMA, 1998), urban sprawl can occur due to two factors: the low density of residential areas. Due to the fact that they have large individual plots and the discontinuity of the same, where it is unused for future use according to their needs.

2.2.2 **URBAN** SPRAWLING

Known as a phenomenon in several cities, Burchell et al. (2003) summarises urban sprawl as a dispersion caused by urban occupation, in places close to rural areas or in spaces far from major city centres.

Polidoro (2012) states that cities have characteristics that demonstrate the presence of urban sprawl, as the urban voids present in various regions. These contribute to the emergence of other urban problems.

The main negative aspects of this phenomenon are presented by Polidoro (2012 apud BURCHELL, 1998):

- Dispersed urbanisation: the creation of allotments in places far from the consolidated centre, creating numerous urban voids;
- Development of areas with low population density: occurs mainly in areas with individual houses (singles) or in those areas with higher incomes, leaving much of the installed infrastructure obsolete or underused; it can also be characterised by locations with only commercial

use.

- Urban expansion area: the disorganised physical-territorial delimitation of urban expansion areas reduces the agricultural area and incites the valorisation of land in the rural-urban transformation process.

Behind these negative aspects there are positive points that end up being drowned out by so many urban problems, Bhatta (2010) highlights the benefits of this growth:

- High economic output;
- Opportunities for the underemployed and the unemployed to have a better life;

Each region, city or state must know its own adversities and make the best use of them so as not to suffer from the impacts of urban sprawl. For this to happen, as already mentioned in the research, a preliminary study must be carried out, so that goals and objectives can be set according to the needs of each urban network.

2.3 MEDIUM-SIZED BRAZILIAN CITIES

Cities are considered to be a phenomenon of agglomerations of large industries, leisure areas, shops, financial and religious centres that satisfy the interests of the population. These agglomerations are so called because each service contained within the cities has a function to fulfil. Each city can be categorised hierarchically according to its economic activity, as its size expresses its main source of development Castells (1983 apud STAMM, 2012).

According to Amorim Filho (1984 apud STAMM, 2012), the first studies on medium-sized cities believed that in order to be considered a medium-sized city, it had to have a number of factors:

- Constant and lasting interactions both with its subordinate regional space and with urban agglomerations of a higher hierarchy;
- Sufficient demographic and functional size to offer a wide range of goods and services to the micro-regional space and to develop the urban economies necessary for the efficient performance of productive activities;
- Ability to receive and retain migrants from smaller towns or rural areas by offering them work opportunities. Thus interrupting the migratory movement towards the big cities, which are already saturated;

- Necessary conditions for establishing dynamic relations with the micro-regional rural area it involves;
- Differentiation of the intra-urban space with a functional centre that is already well individualised and a dynamic periphery like the big cities. Where they are mediated by the multiplication of new peripheral housing centres.

According to Santos (2008), two main factors have contributed to the increase in the urban network: population growth and the accelerated process of urbanisation. The most altered regions in their urban network can be seen in Brazil's large and medium-sized cities. And this distribution leads to an increase in social inequality, whether in economic activities, in the networks that form cities or in the division of the regions that exist in these centres.

The same author explains that medium-sized cities are made up of a real infrastructure that develops over time with the arrival of new inhabitants, absorbing this surplus in the most appropriate way possible. And these can be understood as centres that integrate both national and regional metropolises, as they develop logistical support for large nearby centres.

Ramos (2011) explains that there is a major problem in characterising a medium-sized city, as there are numerous concepts to be understood on this subject. The most common is demographic: according to the IBGE, cities with a population of between 100,000 and 500,000 inhabitants can be considered medium-sized.

Table 02 shows the growth of cities considered to be medium-sized from the 1940s to 2010, with a significant increase between the decades. Comparing the results from 1991 to 2010, the number more than doubled in just 19 years.

Table 2: Brazilian cities with a population between 100 and 500 thousand inhabitants between 1940 and 2010

Year	Number of Cities
1940	08
1970	25
1980	49
1991	113
1996	161
2000	193
2010	245

Source: IBGE Adapted from CONTE (2013).

According to Conte (2013), Brazilian urbanisation underwent major changes from the 20th century onwards, due to the industrial development that drove this progress. Based on the 2000 census, 80 per cent of the population lived in urban areas and 40 per cent of this percentage, almost half of these inhabitants lived in cities considered medium-sized.

These cities can be considered a balancing point due to their urban area. The fact that they are located between large cities and small towns means that they act as intermediaries in some industrial sectors. These areas therefore present opportune locations for investment, and their influence contributes to the development of local industries (CONTE, 2013).

2.3.2 URBAN MOBILITY

For Magagnin et al. (2008), disorganised growth interferes with the whole dynamic of the city and the population's quality of life. It modifies spatial expansion, increases the number of cars and contributes to environmental pollution.

In smaller cities, passers-by can carry out their activities on foot, by bicycle, or in any way that is most pleasant for the user. The big problem in capitals and medium-sized cities is the lack of adequate transport to minimise the time users spend using any means of transport. A survey carried out by IPEA (Institute of Applied Economic Research) in 2010 showed the percentage of the most used means of transport across the country. The centre-west region uses cars the most, with 36.5%, and the northeast region the least, with 13% (Figure 03). When it comes to public transport, the south-eastern region stands out with 50.7% and the north-eastern region uses it the least with 37.5%.

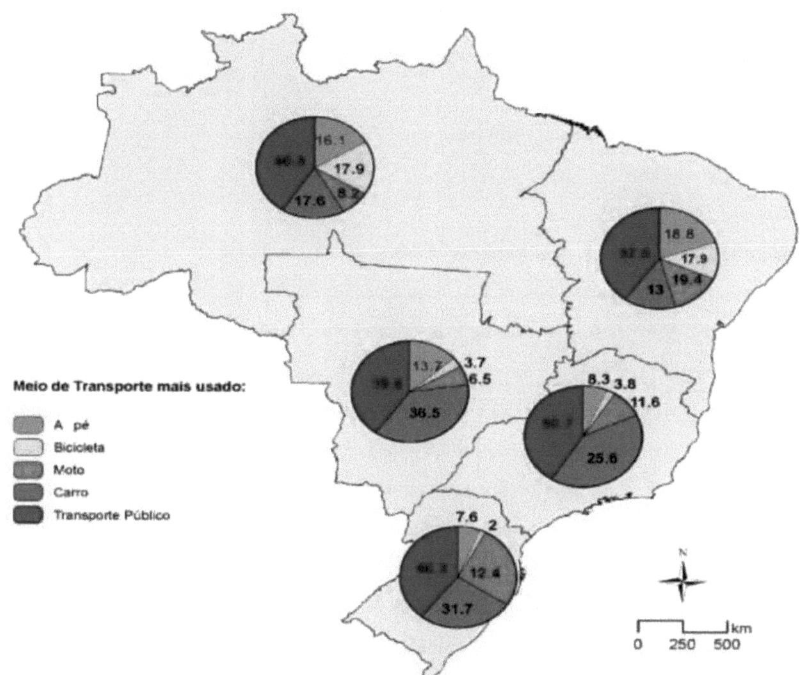

Figure 3: Most used means of transport in Brazil
Source: MAGAGNIN, et.al. (2008).

Urban mobility is one of the subjects most discussed by researchers when it comes to making decisions that minimise and discuss new solutions to urban issues MAGAGNIN et al. (2008). This is because a large part of the population spends most of their time travelling to work and home due to the large flow of traffic and the existence of roads that are unprepared for this type of transport. In a survey carried out by IPEA in 2010, users gave their reasons for using certain types of transport (Table 03).

Table 3: Choice of Mode of Transport by User of Each Mode

	On foot	**Bicycle**	**Car**	**Motorbike**	**Transport**
1°	Healthy	Faster	Faster	Faster	Faster
2°	Faster	Cheaper	Cheaper	Cheaper	Faster
3°	Timetable appropriate	Healthy	Room	Timetable appropriate	The only way transport

Source: IPEA (2010).

The most common reasons given were price and speed of transport, factors that everyone is looking for in their city. Initiatives to implement urban mobility programmes should be studied by public bodies to check the best option for each city, as each one has its own specific characteristics. Magagnin et al. (2008) emphasise that the development of sustainable mobility is a factor that meets the present needs of the urban environment without affecting the development of future generations.

To achieve sustainable urban mobility, Campos (2006) argues that managers should apply certain guidelines to minimise these transport issues:

- Investment in public transport using clean energy;
- Policies to restrict the use of individual transport in already polluted areas;
- Increasing the quality of public transport;
- Implementation of traffic and speed control systems;
- Suitability of freight vehicles;
- Routes and stops;
- Urban comfort: adequate pavements, cycle lanes, safe crossings and tree-lined streets.

Each city has its own peculiarities and these must be worked on individually. The areas to be explored are the characteristics that make them different from each other, for PLANMOB (Urban Mobility Plan) there are different types of cities and they can be defined as:

Industrial cities: These are regions or cities that contain large industries, which account for a large part of the city's economy. They generate heavy traffic (rail and road) with a strong impact on the environment and quality of life. Planning in these cities must minimise these large flows, with restrictions on roads and times of day for heavy vehicles.

Dormitory cities: These are cities close to metropolitan regions or places with a high degree of conurbation. They are known as dormitory towns, being more residential and offering little in the way of services. In this case, services should be geared towards intercity travel, as this population commutes daily.

Tourist cities: These show a different behaviour during specific periods, where they receive a much larger floating population than the existing one. In this case, regions must be prepared for the high demand, which affects their entire infrastructure. Planning must focus on urban traffic and transport.

Historic towns: The main problem with these towns is tourism, as the early formations were unable to cope with the large circulation of motorised vehicles. In addition, they have narrow streets and an urban infrastructure that is outdated for today's times, which could damage the historic elements because they can't cope with this demand. This way, planning must be carried out in projects to adapt these towns in the best way possible, preserving their history.

These are factors that must be studied in order to minimise and contribute to the proper development of more suitable roads and means of transport. If municipal bodies do not provide the necessary conditions for good planning, it is up to them to turn to other spheres of government to subsidise their activities.

2.3.3 ACCESSIBILITY IN CITIES

Lima (1998) explains that accessibility can have various definitions depending on the subject. In a general context, it can be defined as a measure of effort to overcome a spatial separation. It is the opportunity for people to move around to their jobs, studies and essential activities.

This access becomes adequate when the urban environment provides its passers-by with efficient transport and suitable urban spaces. Especially for those with reduced ability, who need special places to carry out their activities normally.

According to Brasil Acessível, the phenomenon of urbanisation has triggered a number of problems that have already been mentioned in cities. And these problems directly affect the people who live in them, demonstrating that they were built without keeping up with the development of human beings and observing their diversity.

When people think of accessibility, they automatically think of people who have problems getting around, whether physically or intellectually. However, the term accessibility encompasses various factors, from the lack of minimum conditions for survival, adequate signposting, to places with roads and streets designed in accordance with current standards.

NBR/ ABNT 9050/2004, which deals with accessibility to buildings, furniture, spaces and urban equipment, aims to establish criteria for the construction and adaptation of buildings, furniture, spaces and urban equipment to accessibility conditions.

However, the difficulty of putting these guidelines into practice in the urban environment becomes a major problem for managers. Small details that prevent this from working gradually increase due to a lack of initiative. In order for everyone to be included, physical and sensory differences and the changes that the human body undergoes as it grows must be respected (ALMEIDA et al., 2013).

2.3.4 THE RELATIONSHIP BETWEEN URBAN MOBILITY AND ACCESSIBILITY

BRACARENSE et al. (2014) state that this global debate is taking place on the proper planning of concepts necessary for the development of a city. At these meetings, they discuss the best way of distributing existing resources, thus realising their objectives.

These two concepts can easily be associated with each other, due to the need for these factors to exist in order for a region's road system to be consistent with its characteristics. Many people do not distinguish between the main functions of mobility and accessibility. John (1981 apud RAIA, 2000) defines these two concepts objectively.

Accessibility: relates to the opportunity that an individual, in a given location, has to take part in a particular activity or series of activities. It relates to the mobility of the individual or type of person, the spatial location of opportunities relative to the individual's starting point, the times when the individual is available to take part in the activities, and the times when the activities are available. Thus, accessibility is related not to behaviour per se, but to the opportunity or potential made available by the transport and land use system for different types of people to carry out their activities.

Mobility: is the ability of an individual or type of person to move around. This involves two components: the first depends on the performance of the transport system, and is affected by where the person is, the time of day and the direction in which they wish to travel; the second component depends on the characteristics of the individual, such as whether they have their own car, are willing to pay for a taxi, bus, train or plane; whether they are aware of the options available to them. In other words, the first element is related to the effectiveness of the transport system in connecting spatially separated locations, and the second element is associated with **"the extent to which"** a given individual or type of person is able to make use of the transport system.

2.4 ACCESSIBILITY INDICES APPLIED TO TRANSPORT

There are several Brazilian organisations studying ways to improve the quality of transport in Brazilian cities. The sustainable urban mobility portal (MOBILIZE) carried out a survey of the largest transport systems in Brazil. Figure 04 shows that the states of São Paulo, Rio de Janeiro and Curitiba are among the largest systems in Brazil; the capital Curitiba does not use trains for urban transport. In 2012, the capital of Paraná stood out for having the highest accessibility index on public transport buses in the country, according to research by Mobilize (2011). Curitiba ended the second half of 2012 with 95.6% accessibility of public transport (AGÊNCIA DE NOTÍCIAS DA PREFEITURA DE CURITIBA, 2012).

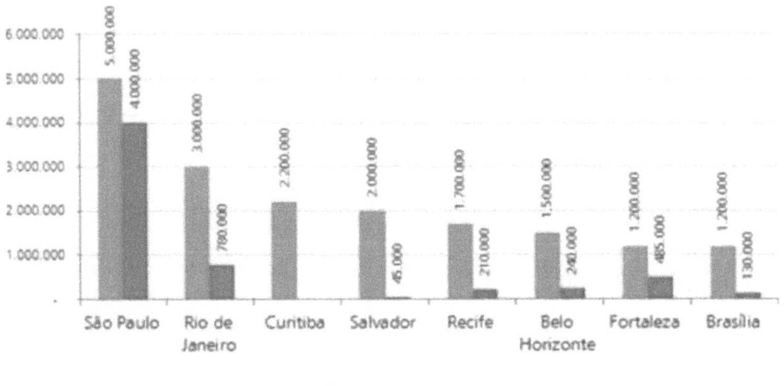

Figure 4: Brazil's largest transport systems (PASS/DIA)

Source: BRACARENSE, et.al. (2014).

When linked to the issue of transport, accessibility is about how easy it is for individuals to get to the right places (BRACARENSE et al., 2014). This same author found definitions for this issue according to various authors, in (Table 01) it is possible to understand some definitions from different perspectives and criticisms and observe the characteristics that lead to this definition.

Table 1 - Definitions of Accessibility by various authors

Features	Definition	Reference
Access from the origin to the transport system	This refers to the user's access to public transport from the place of origin of the journey - home, work, school, etc.	Vasconcellos (2002)
Walking distance from	Micro-accessibility to the transport system can be	Ferraz and Torres

origin to embarkation point	measured in relation to the distance the user walks to public transport, for example between their home and the boarding point	(2004) Santos (2005)
	Users' walking distances from the start of the journey to the boarding point and from the alighting point to the final destination define accessibility to the public transport system.	Ferraz (1998) Andrade et.al. (2004)
Proximity to route to the user's intended activities	0 level of satisfaction among public transport users is directly related to the itinerary of the system offered. The closer the pick-up/drop-off points are to the activities passengers want to do, the less time it takes to make the journey and, therefore, the better the accessibility.	Batista Jr and Senne (2000)
Infrastructure Infra structure transport system and quality	The concept of transport accessibility can be used to assess the infrastructure of the transport system and to locate the regions of a given area or network with inequalities in supply.	Goto (2000)
	Transport accessibility can be indicated in terms of the quality, quantity or user infrastructure offered by the system corresponding to how easy or difficult it is for users to access a desired location within a given area.	Januário (1995) Ferreira (2001)
Localisation of boarding points and frequency	It defines accessibility to public transport from two aspects: locational accessibility - referring to the location of stops close to the origin and destination of journeys, measured in terms of distance or time; and temporal accessibility - indicated by the frequency of routes (lines that serve pick-up/drop-off points and/or terminals).	EBTU (1998)
Physical and economic aspects	Accessibility is considered one domain out of a total of nine for the calculation of a sustainable urban mobility index (SUMI). Within this domain, the topic of accessibility to transport systems is defined in terms of physical access to the network and economic access, subdivided into the following topics: economic accessibility; physical accessibility; and transport for people with special needs - PNE.	Costa (2008)
Efforts to overcome distances	Accessibility as a measure of effort to overcome the spatial separation between two points within an area.	Lima (1998)

Source: BRACARENSE et.al. (2014).

Table 1 presents various concepts related to access in terms of getting around cities, but in relation to this work only LIMA, 1998 addresses this issue in relation to the research. The research was developed on the basis of Lima's research.

Thus Bracarense et al. (2014) state that:

To assess accessibility in relation to an object, it is necessary to use indicators that measure or quantify it. It is preferable to choose indicators that can provide simpler but valid measures that allow for a real and direct assessment.

There are ways of classifying the types of accessibility indicators, and the one that will be used in the research is in the spatial separation group. This is an analysis of the transport network, evaluating the distance travelled and the cost incurred (RAIA, 2000 apud GIANNOPOULOS & BOULOUGARIS, 1989).

2.4.1 ALLEN INDEX

The Allen index is characterised as an indicator of network attributes, of which the road system map is a constituent part, since the smallest intersections or smallest streets can be calculated from the road network. In other words, the shortest distance from point to point of each street and avenue in the urban network is obtained (Dias, 2008).

> These indices are based on graph theory, which studies the relationship between various existing elements of a system made up of sets of points (nodes or vertices), which can be interconnected by lines. When you associate a road network with graph theory, you can say that each zone is represented by a node and the roads by the straight segments that connect them. And this calculation of graph elements allows network properties to be identified. DIAS (2008 apud PIRES, 2000).

This index is classed as an indicator of spatial separation, as it makes it possible to compare regions of the same urban network. According to Lima (1998) the index measures the transposition of overcoming the spatial separation between two points. Equation 01 will be used to analyse the research.

$$A_i = \frac{1}{N-1} \sum_j Dist\ ij \qquad (1)$$

Where :

A_i= Accessibility of intersection i;

Dist ij= Distance between intersections i and j, via the road system;

N = number of intersections used in the calculation.

Souza (2010 apud SANTOS, ZANDONADE; CAMPOS, 2004) reports that these indicators are used for research related to the movement of people, with the aim of understanding the performance of city infrastructure and verifying the results of these levels of economic indicators when applied.

The Allen index is a development indicator that analyses the accessibility of a total area, such as the total road network of a city, and where its neighbourhoods can be analysed by comparing existing subdivisions (SOUZA, 2010 apud LIU, SINGER, 1993).

CHAPTER 3

MATERIAL AND METHODS

This research presents a methodology capable of evaluating the urban sprawl of a city, with the aim of assessing the effects of urban sprawl on the access of passers-by to the urban fabric in a medium-sized Brazilian city

Geographic Information Systems (GIS) and Computer Aided Design (CAD) tools were used to georeference the city's map base. The software used in this research was ARCGIS 10.3 with a licence to use and AutoCAD (Autodesk). The cartographic base of the site under study was provided by the Urban Planning Department of Campo Mourão City Hall, PR, in (dwg)* format and was not georeferenced.

The Technical Registry department of the Department of Urban Planning provided data on all the allotments in the city. This made it possible to separate them by decade in a spreadsheet containing all the allotments with their respective years of approval (Table 04).

Table 4: Year of Approval

Allotments	Year of Approval
Jardim São Sebastião (1956); Centre (1947); Jardim Lar Paraná (1959).	1947 - 1950
Vila Cândida (1963); Jardim Zoraide (1967); Jardim Vitória (1966); Jardim Três Marias (1964); Jardim Tomasi (1967); Jardim Santa Cruz (1965); Jardim Pio XII (1963); Jardim Paraíso do Campo (1968); Jardim Nossa Sr. Aparecida (1963); Jardim Maia I° and 2nd part (1967); Jardim John Kennedy (1966); Jardim Izabel (1963); Jardim Country Club (1968); Jardim Curitiba (1967); Jardim Damasco (1968); Jardim Engenheiro Gutierrez (1969); Jardim Bandeirantes (1963); Jardim Aurora (1968); Jardim Aeroporto (1968);	1960 - 1969
Vila Corinthians (1977); Parque São João (1970); Jardim Tropical (1979); Jardim Sol Nascente (1976); **Joana D'arc Social Garden (1976); Jardim** Indianópolis (1977); Jardim Santa Nilce (1974); Jardim Paulista III part (1979); Jardim Paulista II part (1978); Jardim Paulista (1976); Jardim Modelo (1978); Jardim Maria Barletta (1977); Jardim Lourdes (1976); Jardim Laura (1970); Jardim Florida (1977); Jardim Florida (1978); Jardim Copacabana (1972); Jardim Constantino (1976); Jardim Conrado (1978); Jardim Cidade Nova (1979); Jardim Capricórnio (1977); Jardim Brasília (1973); Jardim	1970 - 1979

Ana Elisa (1977); Jardim Alvorada (1970).	
Industrial Park II (1988); Industrial Park I (1988); Jardim Tropical II part (1981); Jardim Silvana (1983); Jardim Santa Nilce II part (1983); Jardim Orly (1982); Jardim Paulino (1982); Jardim Marino Emer (1980); Jardim Lopes (1987); Jardim Ipê (1984); Jardim Ione (1981); Jardim Fernando (1982); Jardim Araucária (1982); Conjunto Piacentini (1986); Conjunto Parigot de Souza (1987); Conjunto Mario Figueiredo (1988); Conjunto Ilha Bela (1987); Conjunto Milton Luiz Pereira (1980); Parque Residencial Ipê (1987);	1980 - 1989
Condor housing (1997); Jardim Voidelo (1995); Jardim São Luis (1992); Jardim Batel (1997); Jardim Âlcantara (1996); Jardim Albuquerque (1996); Jardim Campos Verde (1996); Conjunto Primavera (1993); Conjunto Montes Claros (1996); Conjunto Mendes (1992); Conjunto Diamante Azul (1993);	1990 - 1999
Jardim Vitória Régia (2002); Parque das Acácias (2006); Jardim Villagio Trombini (2005); Jardim Shangrilá (2004); Jardim Residencial do Lago (2009); Jardim San Marino (2001); Jardim Flora (2003); Jardim Flora II (2009); Jardim Maria Clara (2003); Jardim Flor do Campo (2001); Jardim Cidade Verde (2000); Jardim Cidade Alta (2008); Jardim Casali (2007); Jardim Batel II (2009); Conjunto São Francisco De Assis (2004); Jardim América (2009); Conjunto Avelino Piacentini (2009); Conjunto Governador José Richa (2006).	2000 - 2009
Conjunto Milton De Paula Xavier (2010); Jardim Copacabana II (2010); Jardim Botânico I (2010); Jardim Europa (2010); Residencial Isabela (2011); Residencial Aborê (2013); Novo Centro (2012); Jardim Shangrilá II (2014); Jardim Flor de Lis I (2013); Jardim Flor de Lis II (2013);	2010 - 2015
Conjunto Parque Verde; Jardim Antônio T. Silveira; Jardim Horizonte; Jardim São Pedro	NO YEAR OF APPROVAL

Source: Author, 2015.

Firstly, once they had the (dwg)* file, the unnecessary information was removed and the blocks, plots and information pertinent to the research remained on the base. This was followed up with the use of ARCGIS.

The second stage consisted of using ARCGIS 10.3 software, with the Shapefile (shp*) file separated by decades. This separation was made possible by the data provided by the Department of Urban Planning. Finally, the (dwg)* files were imported into the software for georeferencing. The georeferencing was based on 15 known points in the city analysed so as not to give divergences in the placement of the information on the maps referring to their respective neighbourhoods.

3.1 STAGE OF INTERSECTIONS AND ALLEN INDEX

The Network Analyst extension was used to create the access roads, where a kind of network was created using the Network Dataset command from a shapefile of interconnected lines (Geometric Network), which are the streets in the city's road network.

When a series of network data is built, two new objects are added to this network. Streets, avenues and intersections show the characteristics of the points and all the junctions of the system created since its configuration process.

From these newly generated files, it was possible to handle the new data and thus activate the settings in the software used in the research, where the route to the final objective of the research is developed.

Firstly, a map of street intersections was produced, containing a point centred on the urban perimeter map. This is a meeting point for the other routes to be calculated.

Using the Argcis* software's Data Management Tools >Feature To Point tool, the centre point was located to start processing the data (Map 01). The centre point was used because there was a need for an element as a reference and as the research used LIMA, 1998 as a reference, it used this element.

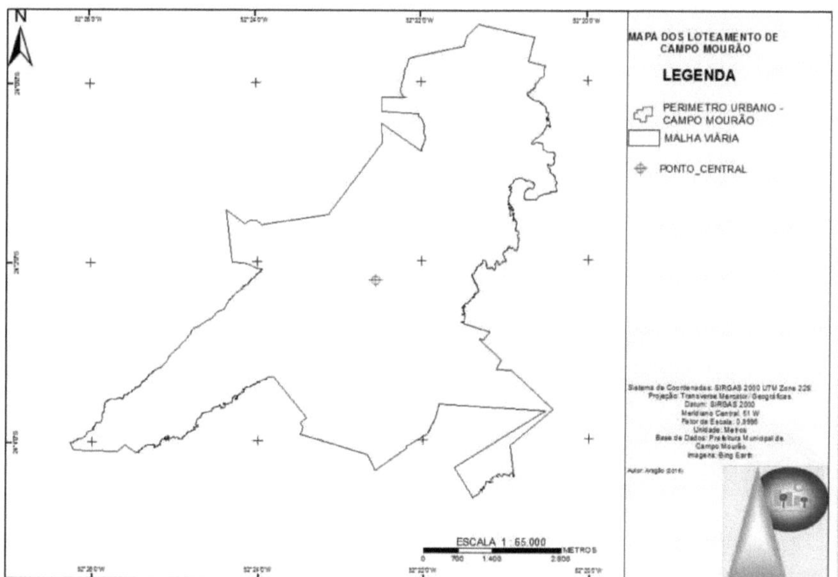

Map 1: Central Point of the Urban Network of Campo Mourão - PR

Source: Author (2016).

In order to mark the points (Map 02), each intersection of existing axes in the urban network was taken.

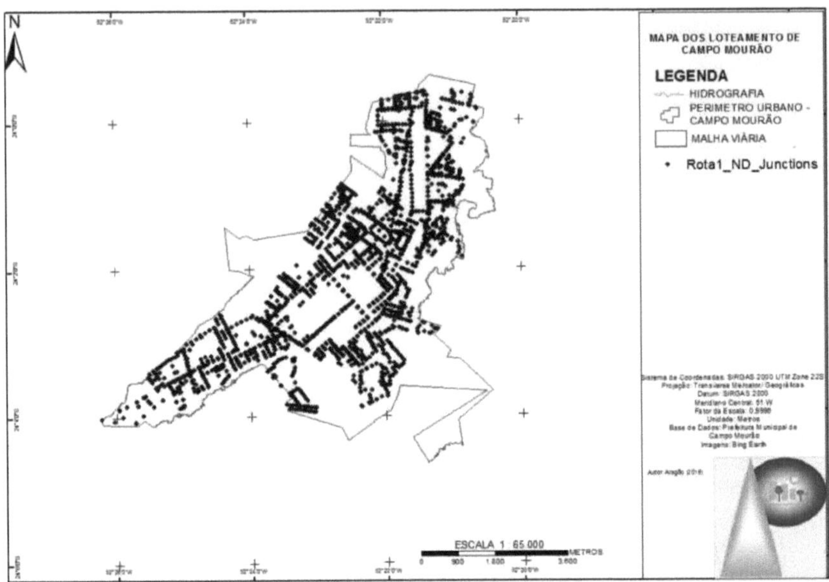

Map 2: Intersection points
Source: Author (2016).

This is how the routes were drawn up (Map 03). Accessibility consists of showing all the accesses to the city being analysed on the route map.

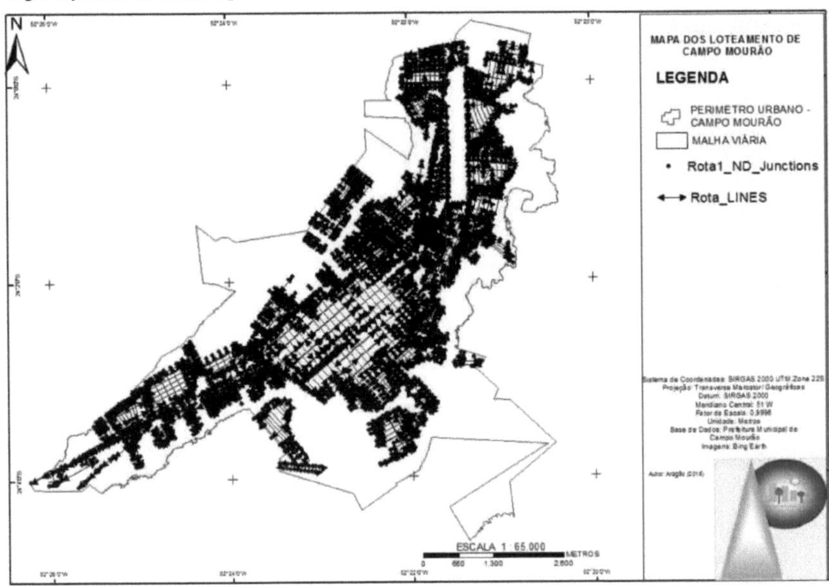

Map 3: Routes delimited by the software
Source: Author (2016).

With the route map open, the next step was to use the Network Analyst command and then the Load Locations tool, selecting a minimum distance of 5000 metres from the central point. The software then calculated the access routes (Figure 05). The measurement of 5000 metres was used as a reference due to the territorial extension of the city for a better understanding of the results.

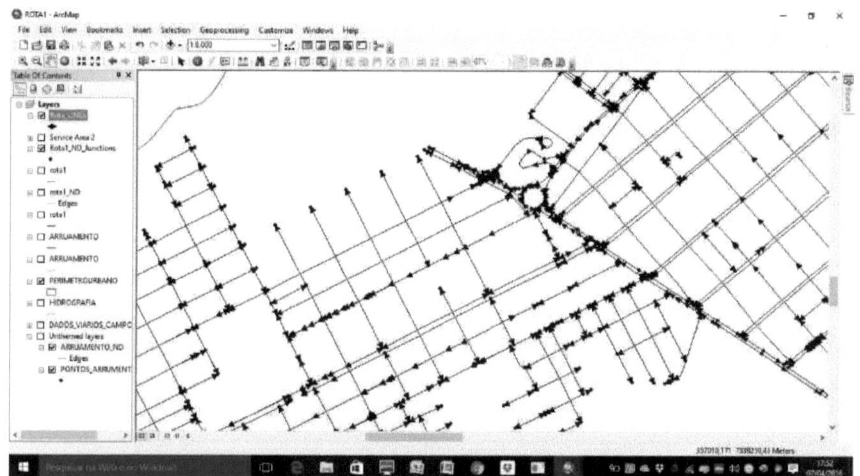

Figure 5: Breakdown of calculated routes
Source: Author (2016).

The routes were calculated by the software, as shown in Figure 05 in detail and clearly demonstrating the routes. To better separate the files, Excel spreadsheets were created, separated by columns for each unknown in the Allen equation to obtain the final results.

After calculating the distances, they were added to the respective column in the excel spreadsheet (Figure 06). Column M (Figure 06) is part of the results obtained by the software, and is also expressed as the sum of the distances from the intersections in Equation 01. The results of the distances from the neighbourhoods to the central point were obtained using the Calculate Geometry* tool, which allows the equation to be calculated and displays the results obtained from the calculated routes.

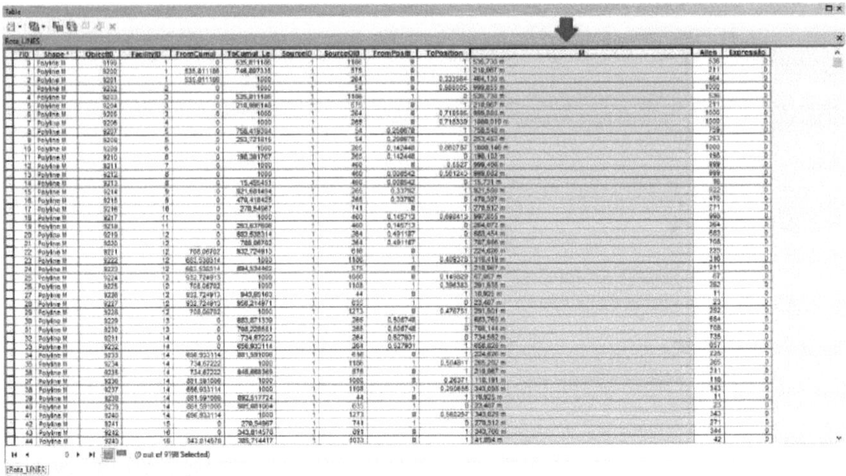

Figure 6: Database
Source: Author (2016).

The database has two more important columns, the expression calculation and the Allen Index. Once you have the expression column labelled M, you need to obtain the values from the Allen column (Equation 04) using the Field Calculator tool to find the desired values.

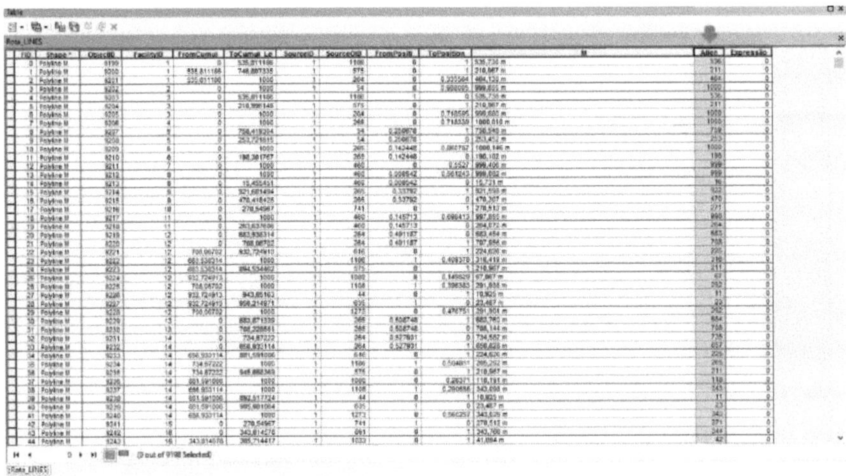

Figure 7: Database Highlighting the Results for Allen's Index
Source: Author (2016).

Once all the columns had been calculated, we proceeded to calculate the index. The next step was to create a file named Allen's General Index, using its formula (Equation 01).

Using a spreadsheet to carry out the calculations, Allen's accessibility indicator was then applied. The information obtained from the calculations was attached to the shp* file, and the ARCGIS software was used to transport the data from the spreadsheet to the database's table of attributes.

Once the table of attributes had been created, the columns were separated into their appropriate values, as shown in Figures 06 and 07.

CHAPTER 4

APPLYING THE METHODOLOGY

4.1 STUDY AREA

The research was carried out in the city of Campo Mourão - PR, to analyse the intra-urban effects of urban sprawl on accessibility in the municipality.

Campo Mourão is located in the north-western region of the state of Paraná (Figure 08), **with a geographical position of latitude 24° 02' 44" S and longitude 52° 22' 59"W.** It is situated at an altitude of 585 metres and has a territorial extension of 763,637 km² (IPARDES, 2015). It was created in 1947 through the emancipation of Pitanga -PR.

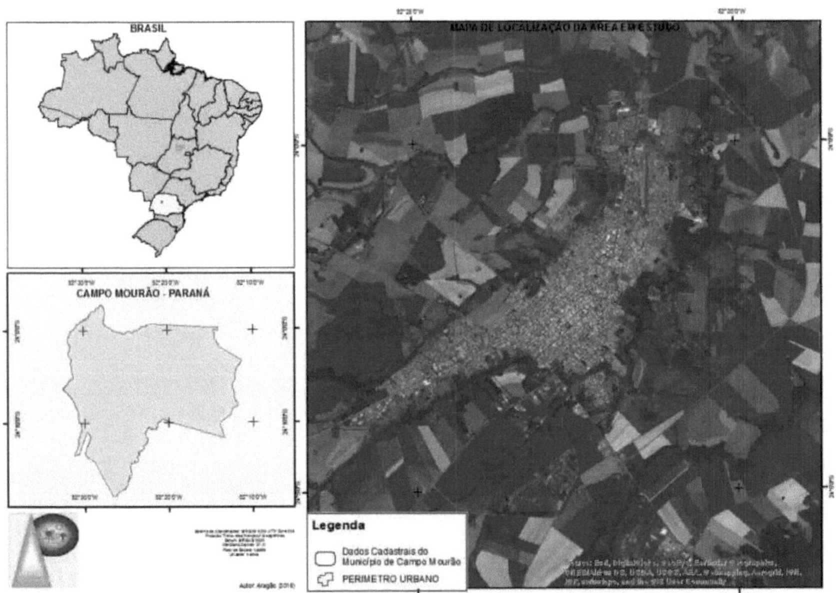

Figure 8: Location of the Municipality of Campo Mourão - PR
Source: Author (2016).

According to Souza et al. (2012), the municipality is located in a region that is economically dominated by modern agriculture. Its main crops are soya and maize and irrigation is not used.

The city borders Araruna, Farol, Luiziana, Peabiru, Mamborê, Corumbataí do Sul and Barbosa Ferraz. The town has two universities, one federal and the other state, and three colleges offering various courses. Due to the arrival of new

The number of students enrolled in courses, neighbouring cities and students from other states end up migrating to Campo Mourão. According to IPARDES, 2015, MEC/INEP reported that in 2013 the number of students enrolled in private, federal and state institutions reached 6,917 students and 942 graduates. This shows that during this period the city gained economically from new residents.

According to the 2010 census, Campo Mourão has an estimated population of 87,194. Of this population, 17.15% are young people aged between 20 and 29 and 15.20% are aged between 30 and 39. Of the total population, 51.80% are female, but based on IBGE data, the estimated population in 2014 is 92,300.

The urban population has a percentage of 94.80 per cent, a small portion of 5.20 per cent is concentrated in rural areas, with a population life expectancy of 75.44 years (Ministério Público Estado do Paraná, 2004).

CHAPTER 5

RESULTS AND DISCUSSIONS

5.1 ROAD SYSTEM

Figure 09 was first drawn up because the road system was then used to calculate all the intersections that have existed since 1947. This is a kind of network in which ARCGIS can calculate the shortest distances from each intersection. These intersections originate from the axes of the meeting of street by street, which are called nodes. The city of Campo Mourão - PR has roundabouts in its urban network and for each street axis that enters the roundabout there is an intersection to be calculated. The purpose of making figure 6 was to check the shortest distance between the intersections

The black lines form the polygons that make up the urbanised or rural areas, but the rural areas were used in the accessibility calculations. Therefore, the results will only be calculated for allotments considered within the urban network.

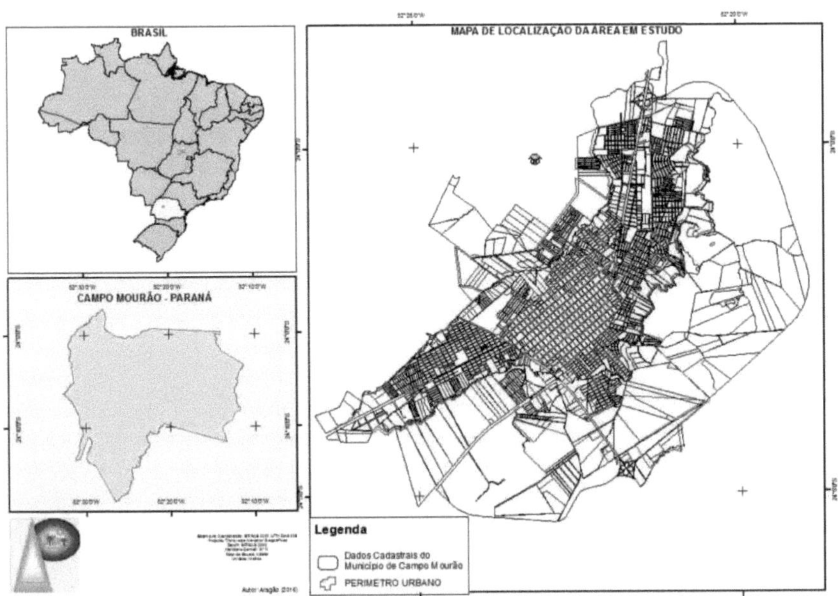

Figure 9: Urban Perimeter and Road System
Source: Author (2016).

Once the road system map had been completed using ARCGIS, a database was created, in which each existing block was given its respective neighbourhood name, plot numbers and block.

The database is made up of the decades of approvals for each subdivision, the number of intersections and the length of the urban network for each decade. The information contained in MAP 04 was gathered by combining all the years of approvals, in which it contains the sites approved from the 1950s to 2014 in highlighted colours . The red part stands out from the rest due to the greater number of approvals in the 60s.

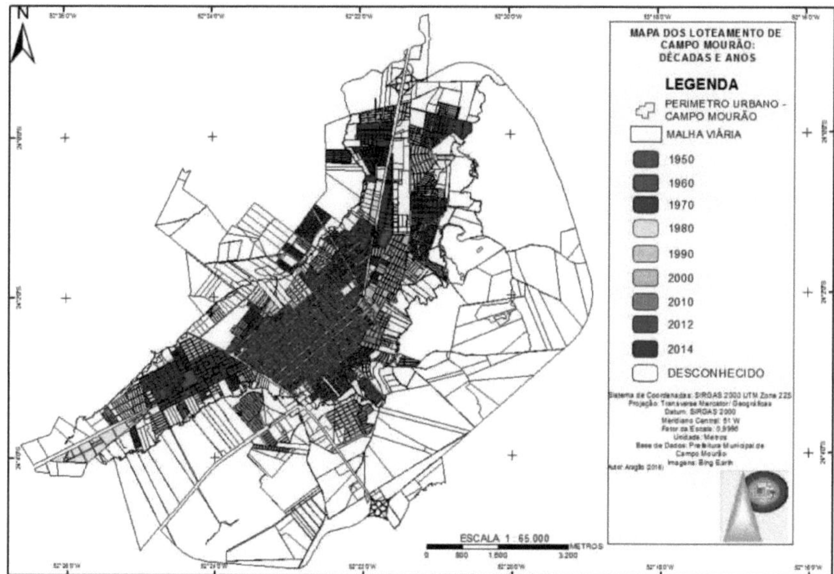

Map 4: Urban Expansion of the Years
Source: Author (2016).

The construction of Map 04 is extremely important for continuing the analysis. Mosaic 01 shows all the allotment developments duly separated by year and decade. You can see which decades and years have seen the greatest growth and which sides have seen the greatest expansion. Mosaic 01 shows the chronological evolution of the growth of Campo Mourão - PR.

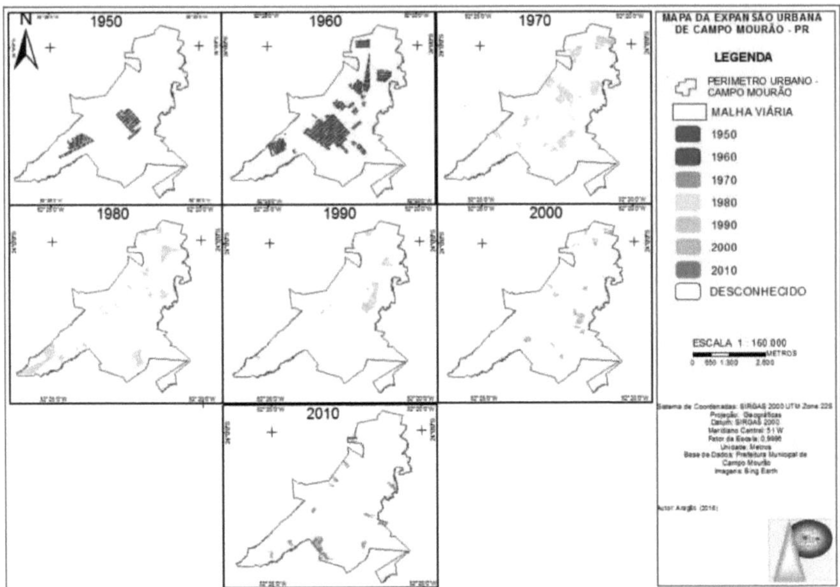

Mosaic 1: Set of Maps

Source: Author (2016).

To better visualise the growth, Mosaic 01 was separated by decade to better describe its evolution. The dispersion is notorious from 2010 to 2014. The largest concentrations occurred from the early 1950s to the 1970s. This dispersion comes from the physical spaces offered by the city's urban fabric.

The extent of the municipality's road network is highlighted in MAP 05, in which the rural extension is in evidence in front of the outline of the urban area

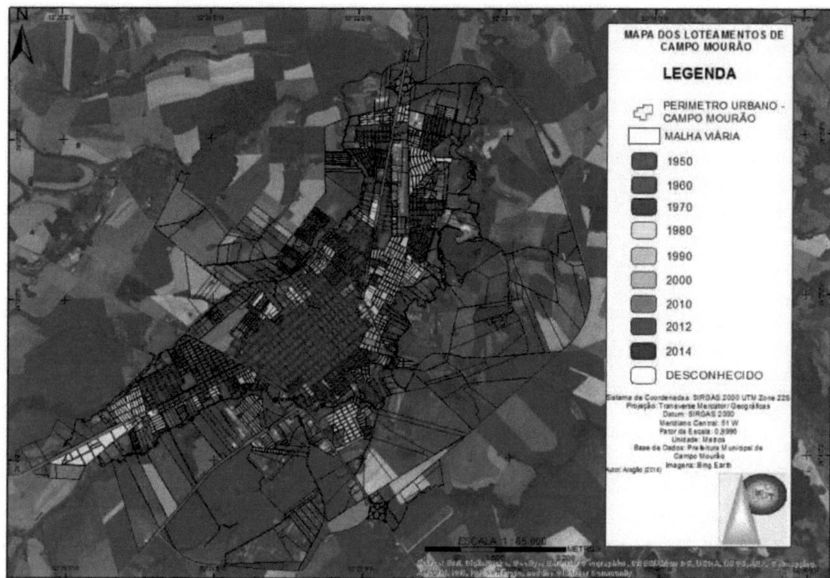

Map 5: Map of allotments highlighting the territorial extension of Campo Mourão - PR
Source: Author (2016).

The intersections stage consists of identifying the number of intersections present in each decade analysed. These intersections are the crossroads of streets on the road network, where two streets meet. This junction was qualified and represented by points on the road network.

In Mosaic 01, the quantity found was separated by their respective decades. This calculation was important for obtaining the routes to the centre point, which was used as a reference for calculating the distance to check the accessibility of each region.

Once the road system map had been drawn up, it was possible to calculate the intersections using the ArcGis tool. The results obtained for the intersections are shown in Table 02, with their respective dates:

Table 2: Intersections of the Decades

DECADES	INTERSECTIONS
1950	247
1960	1009
1970	591
1980	398
1990	279
2000	247
2010	258
TOTAL	**3029**

Source: Author (2016).

Graph 01 shows the number of intersections and shows the decades in which there was the most growth.

Graph 1: Intersections by Decade and Year
Source: Author (2016).

The decades with the lowest intersection values were 1950 and 2000, with 247 intersections, as can be seen in Mosaic 02. In this mosaic, all the decades are separated by their number of intersections. The two decades with the highest number of intersections are 1960 and 1970.

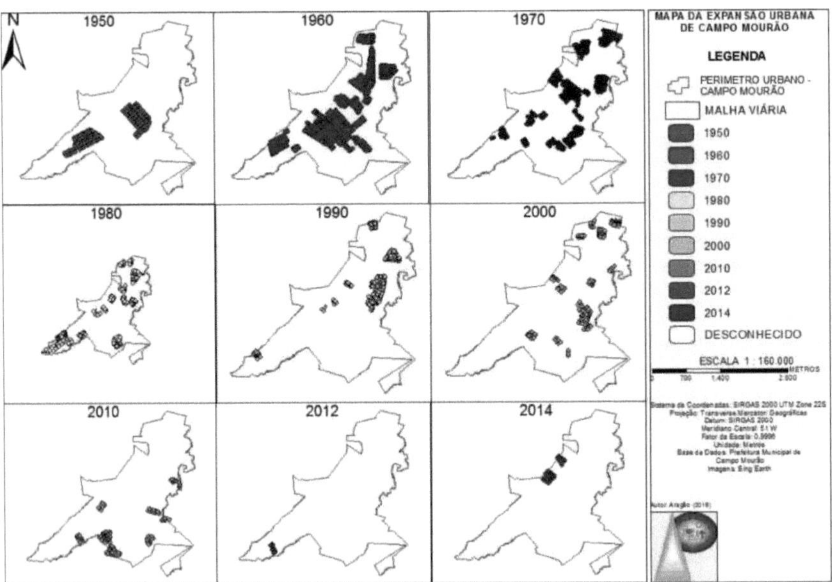

Mosaic 2: Intersections by Decades
Source: Author (2016).

Mosaic 02 is the combination of all the decades that are separated according to MAP 11. The concentration becomes more evident when all are put together, as the city has several intersections. These are fundamental to the accessibility calculation.

5.2 RESULTS OF EXPANSION BY INTERSECTIONS

MAP 06 shows the 1950 expansion with its respective intersections. To obtain these, the ARCGIS tool was used, due to its agility in data processing. The number of intersections found for this decade was 247, the same as for the 2000s. This data shows that densification in the 2000s was relatively low compared to the beginning of colonisation; factors linked to the economy ended up affecting the growth of cities.

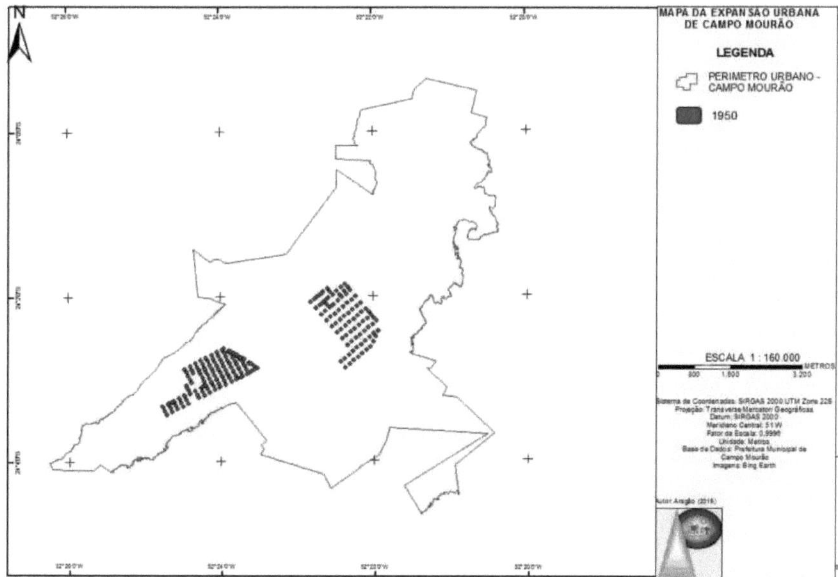

Map 6: Intersection of the 1950s
Source: Author (2016).

The number of intersections that existed in 1960 compared to the 1950s is significant, as it almost tripled to a total of 1009 street junctions. As can be seen on Map 07, their scope is concentrated in almost all regions of Campo Mourão's urban network. With the increase in the number of neighbourhoods, it was necessary to build access roads to reach these places, such as streets, avenues and roundabouts to help the flow of passers-by.

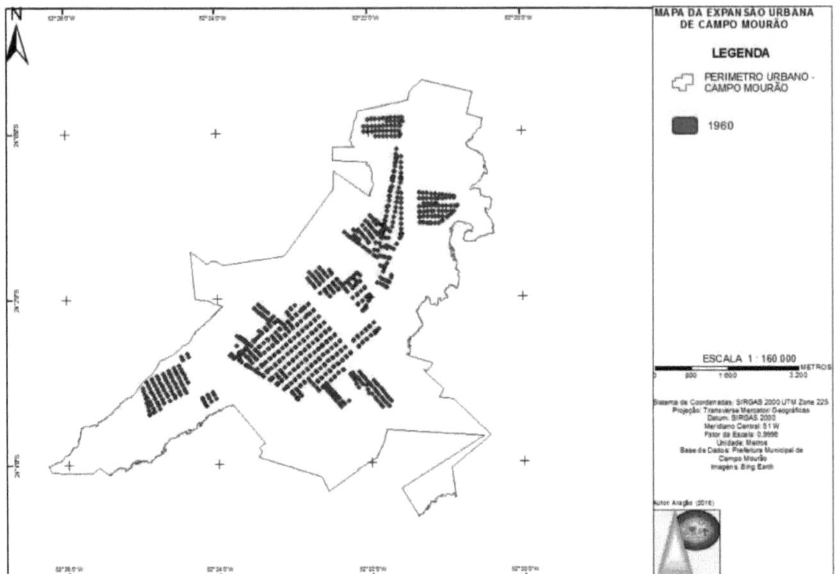

Map 7: 1960s Intersection

Source: Author (2016).

From the 1960s (MAP 07) to the 1970s there was a decrease of 418 intersections, but the spread of approved neighbourhoods in this decade/years compared to the first years became greater. This data can be obtained from the IBGE, as mentioned above.

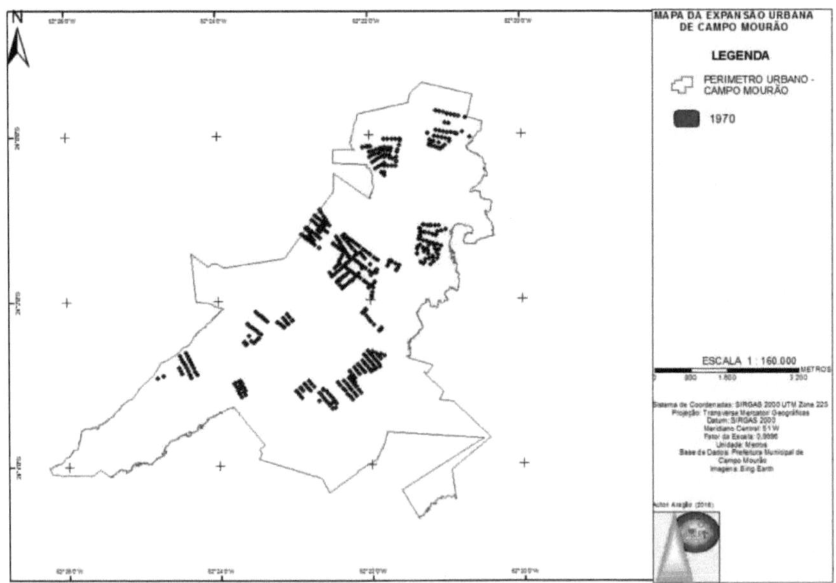

Map 8: Intersection of the 1970s

Source: Author (2016).

The dispersion of the 1980 allotments (Map 09) is considerable, as it covers the south-west, north-east, north-west and south-east regions of Campo Mourão. The number of intersections was low, with a total of 398 intersections. This compares to the previous decade, and especially the 1960s, when allotments were at their peak. Over the years, the interconnection of streets and avenues became necessary for access to various points within the city.

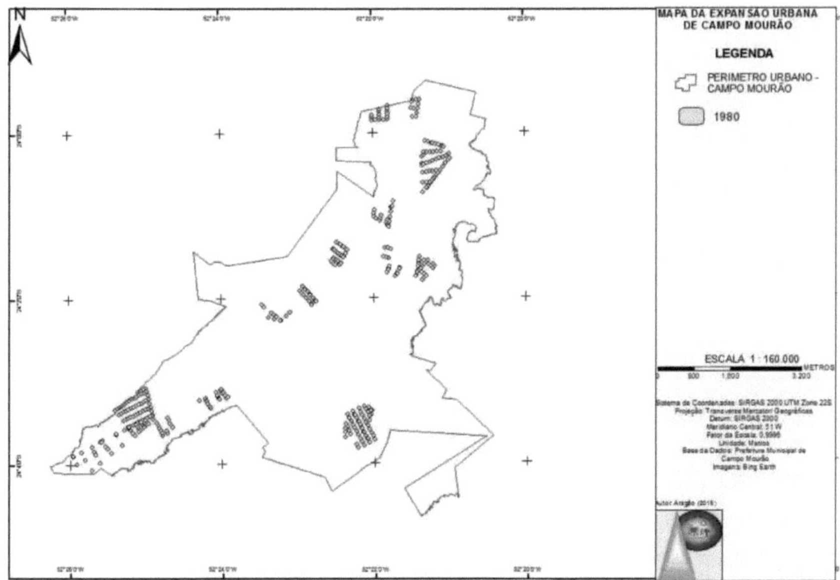

Map 9: Intersections of the 1980s
Source: Author (2016).

In the 1990s (Map 10), the number of allotments decreased, with a total of 279 intersections. Numerous factors can influence this issue, such as the process of approving an allotment, which has become more costly. In addition, city planning has had a direct impact, as there are now regulations for new neighbourhoods. Growth has become concentrated on the edges of the urban network, and only part of it close to the centre.

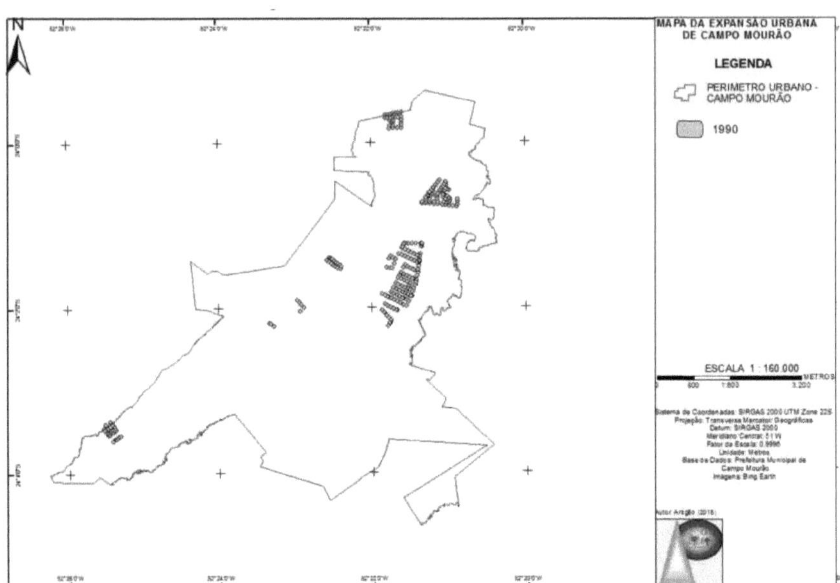

Map 10: Intersection of the 1990s
Source: Author (2016).

The 2000s saw the emergence of 247 new intersections (Map 11). Compared to the previous decade, the difference was only 32 fewer intersections calculated by the GIS tool. Visually, the 1990s had a higher number of intersections due to their agglomeration, which is related to the number of allotments approved.

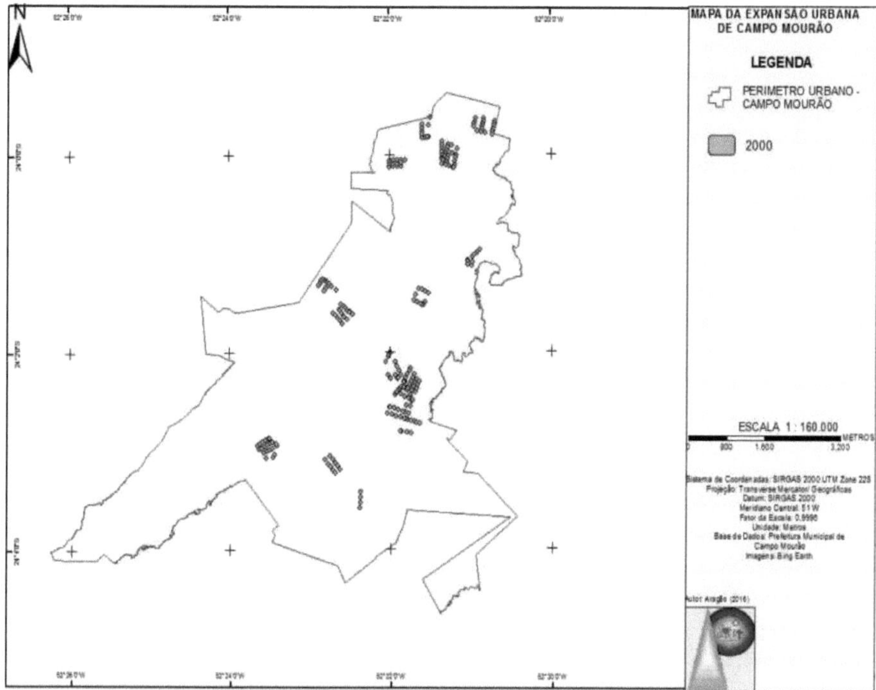

Map 11: Intersection of the 2000s
Source: Author (2016).

With a total of 258 intersections, the 2010s saw an increase of 09 intersections compared to the previous decade, as a result of the approvals and implementation of new neighbourhoods (Map 12). The concentration is more in the south-western region, while the concentration of the 2000s is in the north-eastern region.

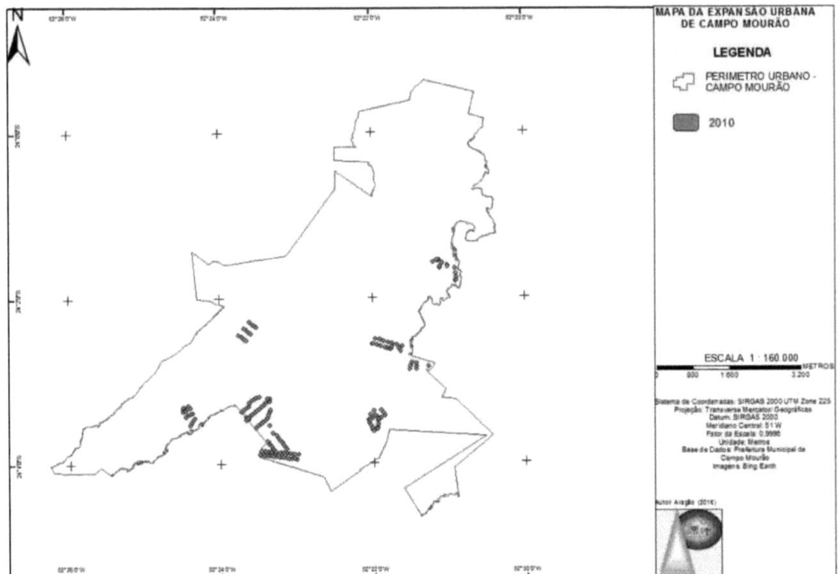

Map 12: Intersection of the 2010s

Source: Author (2016).

Analysing from the beginning of the creation of the city of Campo Mourão, there was growth until the mid-1960s, after which there was a significant drop. Finally, in the 2010s, the numbers began to increase, causing the city to expand further.

CHAPTER 6

ALLEN INDEX RESULTS

To generate Mosaic 03 of the general index by decade, the Network Analyst tool was used with the centre point of the urban perimeter (Map 01), and with the aid of the New Service Area attribute, the map was configured to generate the Allen index according to the distances from the point and the tax table in the accessibility file. In this way, an index was generated for each decade of allotments in the city of Campo Mourão - PR.

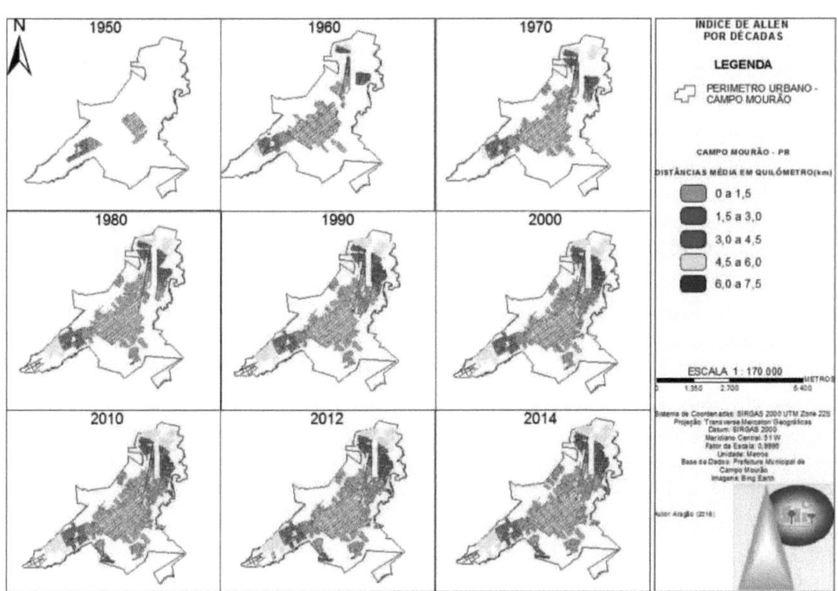

Mosaic 3: Decades Allen Index by Decades and Years
Source: Author (2016).

Mosaic 03 is subdivided into the decades proposed by the study. As the decades grew, the number of streets expanded in proportion to the number of new housing developments. In this way, a file was generated for the index maps, which presents several shapefiles separated by decades.

It is thus subdivided into a general index shapefile with street distances from 0 to 1.5 km; 1.5 to 3.0 km; 3.0 to 4.5 km; 4.5 to 6.0 km and 6.0 to 7.5 km. The choice of these distances refers to the research carried out by Lima (1998). However, it was decided to find a distance of up to 7.5 km due to the territorial extension of the city. If other studies choose to analyse larger cities, they can continue to establish larger distances as Lima, 1998 used.

In the 1950s the streets did not have the maximum distances analysed of 4.5 to 6.0 km and 6.0 to 7.5 km. This means that at that time there weren't so many mobility and accessibility problems for users (Map 15).

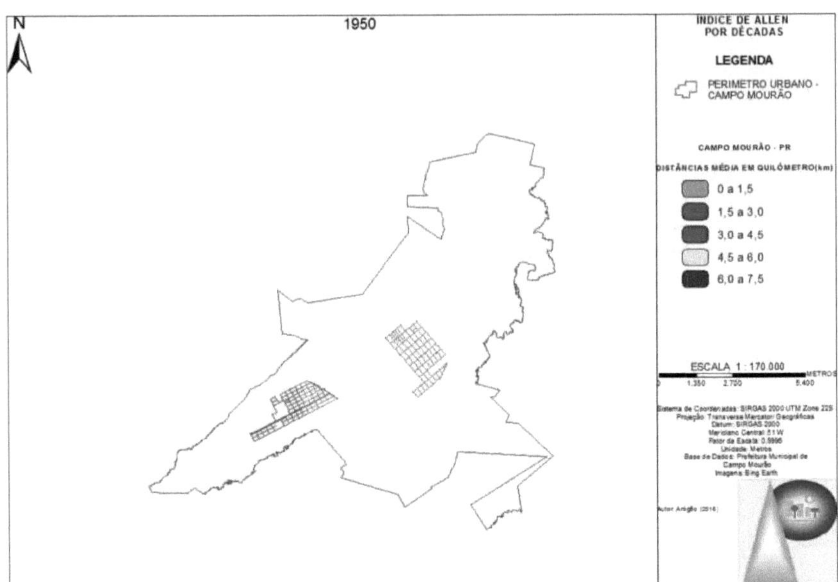

Map 13: Accessibility in the 1950s
Source: Author (2016).

Map 16 from the 1960s shows a great evolution compared to the previous one. The number of neighbourhoods grew and, consequently, the number of streets increased. From this stage of the analysis, the proposed distances were calculated. The lines in light blue correspond to the central region of Campo Mourão - PR, these are the distances of up to 1.5 kilometres according to the map legend, and they showed the highest concentration of streets.

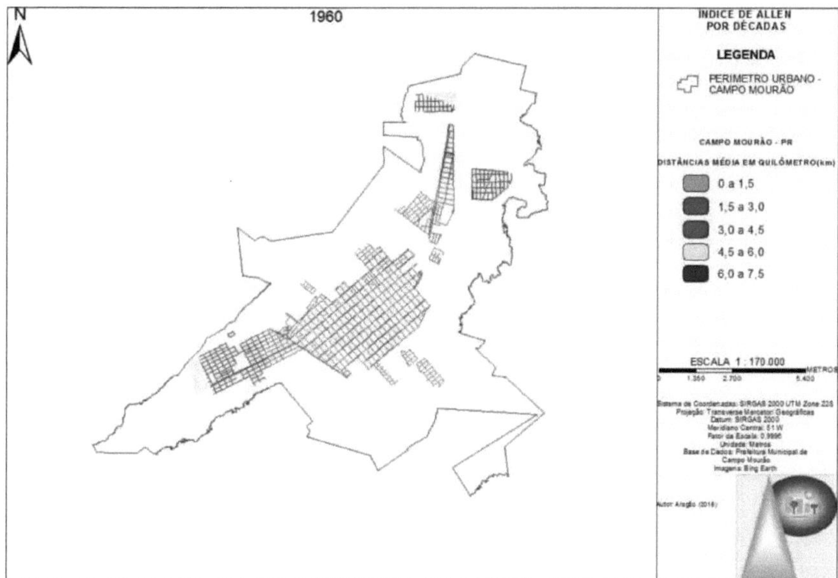

Map 14: Accessibility in the 1960s
Source: Author (2016).

From the 1970s onwards, the number of streets increased (Map 27). As the index is calculated from the centre point of the urban perimeter, the distances have consequently increased. As such, the shortest distance established in the analysis of 0 to 1.5 km is not found on the 1970s map, as the software reads the longest distances so the shortest distances are not calculated, as the aim was to identify the longest routes.

From distances of 1.5 to 3.0 kilometres, the concentration of streets was greater than the others, Map 17. The extremities of the city map showed greater distances and lower accessibility. This indicates that these regions have longer journeys to reach the city centre, making accessibility more difficult.

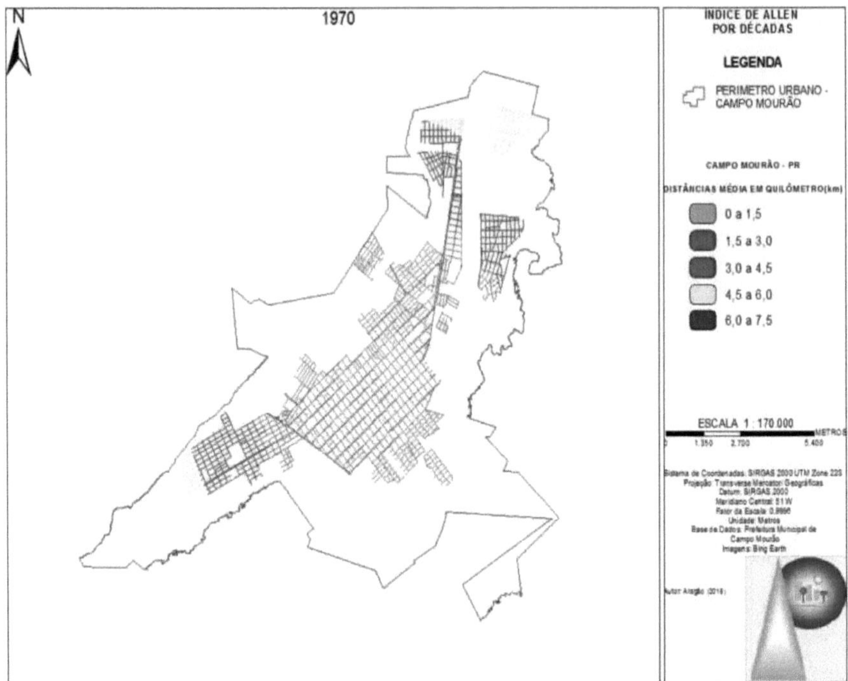

Map 15: Accessibility in the 1970s
Source: Author (2016).

The great growth dividing the decades was from 1970 onwards, so accessibility at some points became less. Map 18 from the 1980s shows some of the same characteristics as the previous decade, but the greatest distances from the centre point of 6.0 to 7.5 km appeared in the western region.

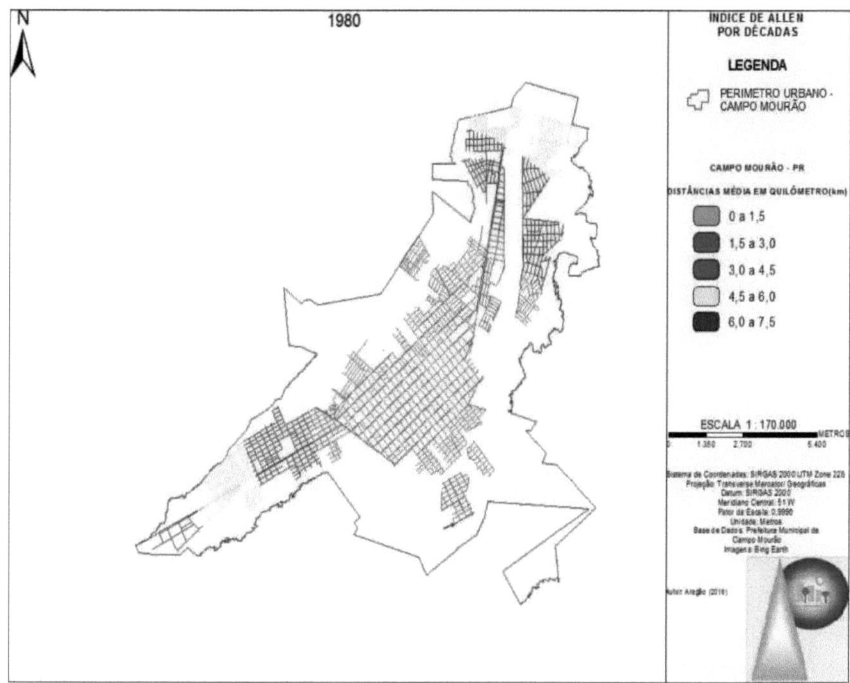

Map 16: Accessibility in the 1980s
Source: Author (2016).

As allotments have increased, the distances have followed this development, and with it the accessibility has become less. Thus, the shorter the distance, the greater the accessibility. The north-east/south-west regions have greater distances, with the west region standing out, with stretches of 6 to 7.5 kilometres.

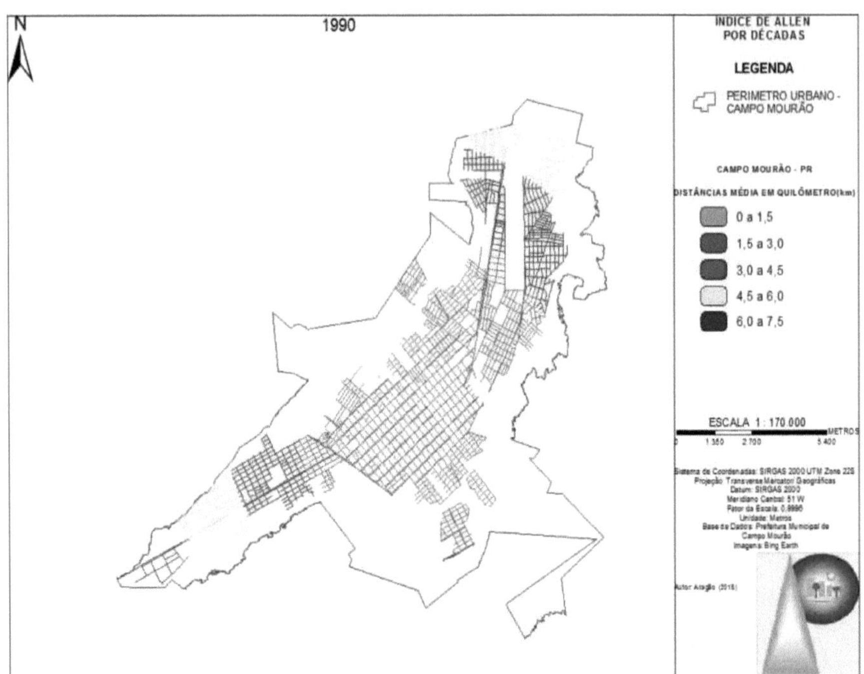

Map 17: Accessibility in the 1990s
Source: Author (2016).

In the 1990s, the streets with the greatest accessibility were at distances of 1.5 to 3.0 km. Map 19 shows the green region, which predominates in the central region. The distance of 3.0 kilometres was present in the majority of streets from the 1970 maps onwards.

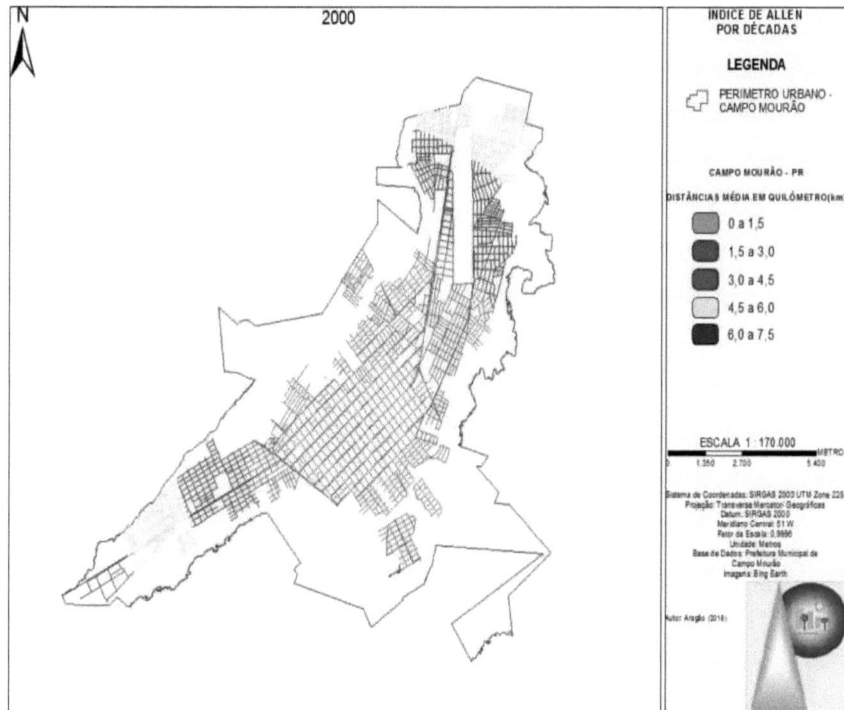

Map 18: Accessibility in the 2000s
Source: Author (2016).

The shape of the city's urban grid influences accessibility, as it can be seen that the streets (Map 20), which are coloured red, are on the edges of the southwest/northeast region of the city. These calculations were based on the centre point used as a reference for calculating all the existing streets.

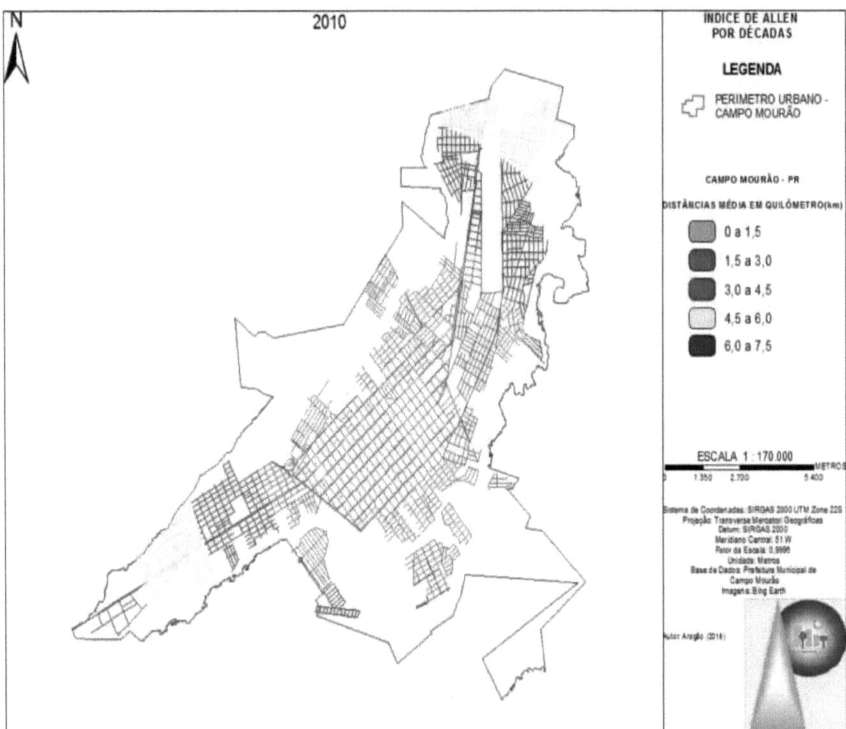

Map 19: Accessibility in the 2010s
Source: Author (2016).

The extremities of the city have greater distances of 3.0 to 4.5 km and 4.5 to 6.0 km, revealing the difficulty in accessing the centre (Map 21). This implies the proper development of the urban fabric and its growth.

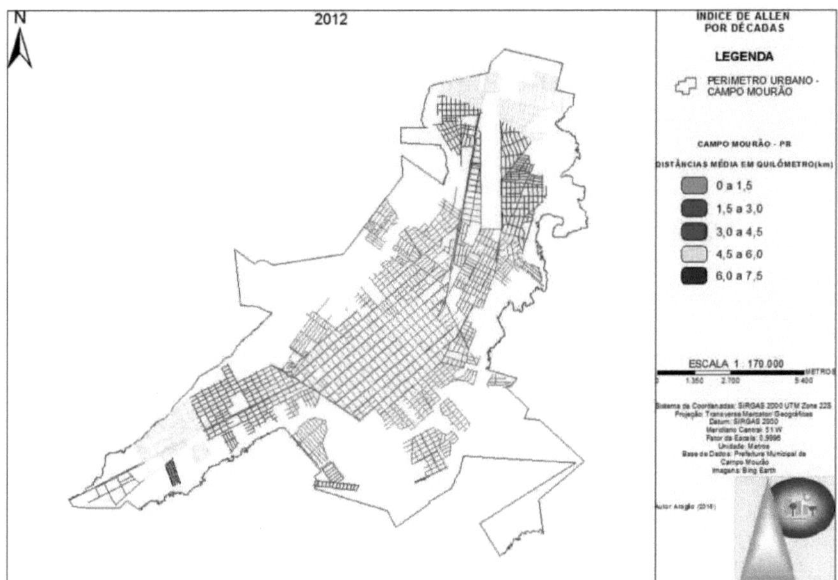

Map 20: Accessibility in 2012
Source: Author (2016).

Although the growth of new allotments has been significant since 2012, their streets continue to follow the same criteria as the others from the 1970s onwards (Map 22). Since the areas where new allotments have been set up are located on the edges of cities, there is a tendency to increase the distance as they expand.

Map 23 for 2014 differs little from the previous year analysed. This one shows the presence of greater distances due to the new neighbourhoods. However, due to municipal progress, the urban network has grown significantly, as can be seen in Map 23.

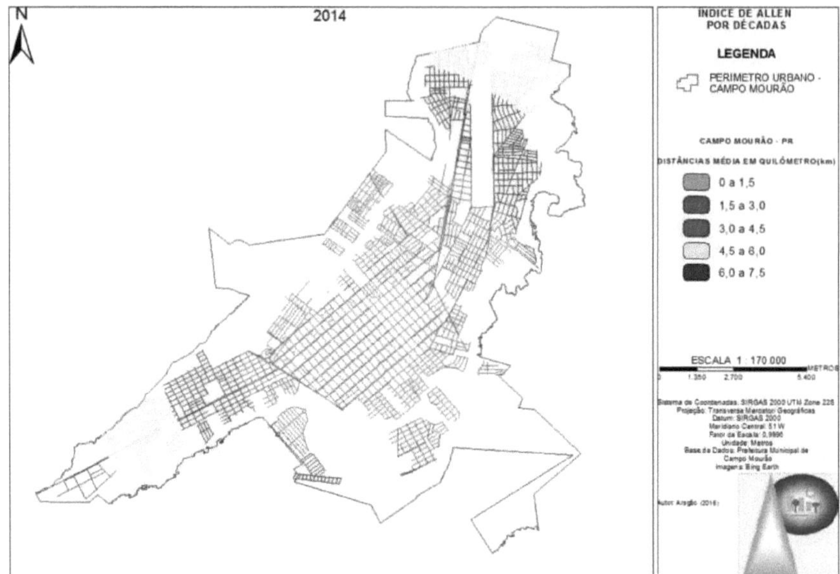

Map 21: Accessibility in 2014
Source: Author (2016).

By calculating the shortest and longest distances, we can discover the influence of the different shapes on accessibility. This is because the shape of a city's growth interferes with its development. To this end, the results of the number of intersections will be evaluated according to the distances established in the calculations for all the decades analysed.

3.2 VALUES RESULTING FROM THE APPLICATION OF THE ALLEN INDEX

The results obtained by analysing accessibility by decade are shown in Tables 7, 8 and 9. Table 7 shows the number of intersections found by the software by distance interval and the percentage found.

In the 1950s, the number of intersections with a 1.5-kilometre lane was 38.05%. As the distances increased, the number of intersections decreased, reaching zero in the last 7.5-kilometre lane. The 3.0 to 4.5 kilometre lane was 27.12% lower than the first lane.

According to LIMA (1998), in his research there was an increase in the percentage of intersections, where the 1.5 kilometre lane had 24% while the next 3.0 kilometre lane had 76% of intersections. If we compare the two studies, in the first decade analysed by Lima there was an increase in the first two lanes and the following lanes had zero intersections. In the first decade in Campo Mourão, PR, there was a decrease from the 1.5 km to 4.5 km lane, and the remaining lanes also had zero intersections.

In the 1960s, the first 1.5 kilometre strip showed 38.55% and the next three showed a decrease in percentage, with only the 6.0 to 7.5 kilometre strip showing no intersections. The 1960s, according to LIMA (1998) , behaved in the same way as the proposed study from the second lane onwards: while the first lanes showed percentages, the last had zero intersections. This shows that the greater the expansion, the lower the accessibility.

The 1970s began differently, with the first lane having no intersections. After the second stretch from 1.5 km to 3.0 km, the intersections decreased, and in the last stretch there were no intersections at all. In this decade, the results were the same as Lima's 1970 table, reinforcing the lack of accessibility when urban growth becomes progressive.

Table 5: Allen's Accessibility by Decade

	Until 1950		Until 1960	
Categories (Km)	Intersections	%	Intersections	%
0,0 a 1,5	94	38,057	389	38,553
1,5 a 3,0	86	34,818	375	37,166
3,0 a 4,5	67	27,126	225	22,299
4,5 a 6,0	0	0	20	1,982
6,0 a 7,5	0	0	0	0
Total	247	100	1009	100
Index E (Km)	1,71		0,41	

Source: Author (2016).

Of the first three decades analysed, the second showed the greatest urban growth and development. From the 1980s onwards (table 6), the first strip did not intersect. This means that the spaces close to the centre were all occupied and growth took place in distant places due to the city's needs. However, in this decade, the 4.5 km to 6.0 km range saw an increase in intersections, with 30.15 per cent of the percentage, lower than the second range, which showed 46.73 per cent. There were no results for the first lane alone (Table 05), the same as Lima's research, which used the Allen Index. Allen can carry out a spatial separation analysis, from neighbourhoods to different decades. That's why the study of suitable sites with adequate infrastructure is essential.

Table 6: Allen's Accessibility by Decade

	Until 1970		Until 1980	
Categories (Km)	Intersections	%	Intersections	%
0,0 a 1,5	0	0	0	0
1,5 a 3,0	453	76,65	186	46,734
3,0 a 4,5	102	17,259	82	20,603
4,5 a 6,0	36	6,091	120	30,151
6,0 a 7,5	0	0	10	2,512
Total	591	100	398	100
Index E (Km)	0,7 1		1,06	

Source: Author (2016).

LIMA, 1998 observes that the trend continues for the other decades of his research, where the values of the bands are distributed among the higher bands. In the 2000s, the range from 4.5 km to 6.0 km was 30.15% and the second range from 1.5 km to 3.0 km 46.73% (Table 07). From the 1960s to the 1980s, the 6.0 km to 7.5 km lanes had intersections.

Table 7: Allen's Accessibility by Decade

	Until 1990		Up to 2000	
Categories (Km)	Intersections	%	Intersections	%
0,0 a 1,5	0	0	0	0
1,5 a 3,0	158	56,631	161	65,182
3,0 a 4,5	66	23,656	33	13,36

4,5 a 6,0	55	19,713	53	21,457
6,0 a 7,5	0	0	0	0
Total	279	100	247	100
Index E (Km)	1,51		1,71	

Source: Author (2016).

As the city grew, the intersections were grouped together near the central point and the others were built along with the new housing developments so that access to the central point could take place.

In the 2010s, only the 1.5 km to 3.0 km and 3.0 km to 4.5 km lanes had intersections. The second had 68.96 per cent, while the third had 31.03 per cent.

The reflection of urban planning has an impact on the figures found in the survey. Only the 4.5 km to 6.0 km interval had 13 intersections in the 2010s, while the other intervals had no intersections (Table 08). This means that there were few allotments set up, so access to the central point from these new neighbourhoods was lower due to the low percentage of growth at that time.

There were no intersections in the first interval because these neighbourhoods were far from the central point. From 2010 onwards, there was an increase of 11 intersections compared to the 2000s.

Table 8: Allen's Accessibility by Decade and Year

	Until 2010	
Categories (Km)	Intersections	%
0,0 a 1,5	0	0
1,5 a 3,0	191	70,479
3,0 a 4,5	54	19,926
4,5 a 6,0	13	4,797
6,0 a 7,5	0	0
Total	271	100
Index E (Km)	1,47	

Source: Author (2016).

Unlike the 2000s, which had the lowest number of intersections in the 1.5 to 3.0 kilometre range when compared to the 2010s in this same range. In the latter year, the intersections found were at a shorter distance than in the previous year, which means that allotments were set up close to the city centre. The city's growth is due to the availability of land and the growth factor of the chosen location.

The global index (E) determines the level of accessibility for each decade analysed. These values were found using the Allen index formula, in which this index determines the levels of accessibility resulting from the urban expansion of cities.

Tables 05 and 06 show the overall indices for 1950, 1960 and 1970. Initially, the intersection with the best accessibility was 0.41 kilometres in the 1960s. It showed the greatest urban growth from the creation of until the last year analysed. The year 1950 was in 4th place along with the year 2000.

The greatest densification was in the 1960s, and over the years there was a decline in the number of new neighbourhoods until the end of the 1990s. However, these rates were not as significant as the years studied from 2000 onwards (Table 07).

There was moderate growth in the 2000s, with the intersection with the lowest accessibility in 2000 (1.71 km) equalling the result from the 1950s. Of the two periods shown in table 06, the intersections in the 1980s had the greatest accessibility.

Thus, in line with the initial hypothesis based on the effects of urban growth, LIMA (1998) confirms, using the Allen Index, that disorderly expansion reduces accessibility levels.

These results are valid for the city of Campo Mourão - PR, as the Allen Index as well as another accessibility indicator called the Davidson Index show similar results when research is carried out in medium-sized cities, as was the case for this study (RAIA, 2000).

This index is able to make a comparison of spatial separation between regions, and was analysed from the 1950s to 2010. In this case, the average distances of the allotments set up over the decades were checked and which had the greatest accessibility in relation to the distances established as a parameter for analysis.

The intersections thus had accessibility of 35.15 km. In the 2010s, the overall index generated was 1.47 km (table 08), the third lowest value of all the periods analysed for accessibility to the city centre.

CHAPTER 7

FINAL CONSIDERATIONS

This study demonstrated the possibility of using the Allen Index, an indicator of spatial separation, which makes it easy to read the results. In addition, the way in which it was applied to the city of Campo Mourão - PR made it possible to verify whether the tool is valid in medium-sized cities.

This methodology enabled the development of thematic maps necessary for municipal planning, serving as a basis for monitoring the growth of the city in question. In this way, managers can use a tool capable of detecting possible improvements that prioritise the quality of life of their users, as well as introducing them into the development process.

The results for access were categorised into distance intervals due to the size of the city. In this way, the results analysed by decades and years were lower as new neighbourhoods were built. The disorderly dispersal of new neighbourhoods also reflected in the results. At times when the neighbourhoods established had distances close to the central point, they showed an adequate overall index when compared to the others.

With regard to users, the results show that access to the central area is less accessible due to the greater distance. The most distant regions are those implemented in the last periods analysed, because the available areas are in distant locations.

It is therefore up to managers to implement basic services that meet the minimum needs of their users. In addition to prioritising neighbourhoods that need greater mobility due to the large flow of people.

The methodology therefore proved to be flexible and can be used in medium-sized cities, as was the area of application. The spatial separation tool proved to be economical, as the form of analysis can be carried out by various planning professionals.

To do this, it is necessary to adapt to each urban space, as well as prioritising the aspects that negatively affect its population. Thus, the verification of urban expansion using this indicator proved to be valid due to its efficient application, as it contributes to comfort, safety and economy because it is an easy-to-apply tool.

REFERENCES

ALLEN, B. W.; **Accessibility Measures of U.S. Metropolitan Areas**. 1993, Transpn. Res- B. Vol. 278. n. 6, p.439 - 449. Pcrtmmon Press Ltd.

ALMEIDA, P.E.; GIACOMINI, B. L.; BORTOLUZZI, G. M.; Mobility and urban accessibility. 2013, p. 1-7. 2nd SNCS - National Seminar on Sustainable Buildings.

ARAÚJO, G. J.; HADLICH, M. G.; ASSUMPÇÃO, P. C. H.; Urban expansion of Jacobina, Bahia, from 1969 to 2008. Anais XVI Simpósio Brasileiro de Sensoriamento Remoto - SBSR, Foz do Iguaçu, PR, Brazil, 13 to 18 April 2013, INPE.

BARIQUELLO, P.M.L.; Geotechnology applied to analysing the urban expansion of Botucatu - SP (1962-2010). 2011. p.141 Dissertation (Master's Degree in Agronomy - Energy in Agriculture). Universidade Estadual Paulista Júlio Mesquita Filho, Botucatu, 2011.

BATISTA, R.M.; CORDOVIL, S.C.F.; Urban and morphological development of Campo Mourão, Paraná, Brazil.

Geoingá: Journal of the Postgraduate Programme in Geography. Maringá, v. 4, n. 2 , p. 77-92, 2012.

CUNHA, B.A.; Process of urban space formation in the municipality of Nobres - MT. 2011. p. 147 Dissertation (Master's in Geography). Federal University of Mato Grosso, Cuiabá, 2011.

CONTE, H.C.; Medium-sized cities: Discussing the theme. 2013, p.45-61. Sociedade e Território, Natal, v.25, n° 1.

COSTA, R.F.R.; LIMA, S.F.; SILVA, O.D.; Local fiscal policy and economic growth: A panel study for northeastern municipalities. v. 44, n.1, p.93-112, 2013.

CRISPIM, Q.J.; MALYSZ, T.S.; CARDOSO, O.; JUNIOR, P.N.S.; Conservation and protection of springs through soil cement on small farms in the Rio do Campo basin in the municipality of Campo Mourão - PR. Revista Geonorte, Special Issue, v.3, n.4, p. 781-790, 2012.

CITY STATUS, Brasilia, 2004.

FRANÇA, A.; Spatial performance indicators case study: The city of curitibanos - SC. 2004, 136p. Dissertation (Master's in Urban and Regional Planning). Federal University of Rio Grande do Sul, Porto Alegre, 2004.

GONÇALVES, R.A.; Indicators of urban dispersion. 2011, 113p. Dissertation (Master's in Urban and Regional Planning). Federal University of Rio Grande do Sul, Porto Alegre, 2011.

IPARDES (Paraná Institute for Economic and Social Development). Statistical Notebook, State of Paraná, 2015.

JUNIOR QUADROS, R.H.; Between buses and cars: The question of prioritising public transport in Brazilian urban mobility. 2011, 165p. Dissertation (Master's in Civil Engineering). Federal University of Santa Catarina, Florianópolis, 2011.

LIMA, R.S.; Urban expansion and accessibility - the case of medium-sized Brazilian cities. 1998, 91p. Dissertation (Master's in Transport). University of São Paulo, São Carlos School of Engineering. São Carlos, 1998.

MAGAGNIN, C.R.; SILVA, R.N.A.; The expert's perception of urban mobility. Transportes, v. XVI, n. 1, p. 25-35, June 2008.

MARIA, R.P.; LUIZ, S.M.; LEIGE, J.L.; CARLOS, J.L.; ALMEIDA, I.; SPATIAL DYNAMICS OF THE POPULATION OF CAMPO MOURÃO - PR. Urban Studies Symposium: Regional Development and Environmental Dynamics. Campo Mourão, 2011.

MORIGI, B.J.; MORIGI, C.M.; Territorial occupation and the evolution of space in Campo Mourão - PR. SEURB II Symposium on Urban Studies: The dynamics of cities and the production of space, p.27, 2013.

OLIVEIRA, N.; BARCELLOS, M.T.; BARROS, C.; RABELO, M.M.; Urban voids in Porto Alegre - Capitalist land use and social implications. Secretaria e Coordenação e Planejamento. v.6, p.94, 1991.

POLIDORO, M.; Conurbation and dispersion in urban agglomerations: Challenges for planning. 2012, 205p. Dissertation (Master's in Urban Engineering). Federal University of São Carlos, São Carlos, 2012.

POLIDORO, M.; Reflections on the influence of the BR-369 motorway in defining the urban expansion patterns of municipalities in the state of Paraná. 2011, p.1-8. IX ENANPEGE - National Meeting of the National Association for Graduate Studies and Research in Geography.

PEREIRA, M.C.J.; Importance and significance of medium-sized cities in Amazonia - An approach from Santarém - PA.

2004, p. 139 Dissertation (Master's Degree in Sustainable Development of the Humid Tropics from the Centre for High Amazonian Studies). Federal University of Pará, Belém, 2004.

RAMOS, F.E.; Medium-sized cities in perspective. An analysis of the roles and trends of a group of medium-sized Brazilian cities.2011. p.164 Dissertation (Master's in Geography). Federal University of Minas Gerais, Belo Horizonte, 2011.

ROSOLÉM, P.N. Cartographic visualisation of the expansion of the city of Londrina through a collection of digital maps. Ambiência Guarapuava (PR) v.8 Ed. Especial - 1 p. 667684. 2012.

STAMM, C.; STADUTO R.A.J.; LIMA, F.J.; WADI, M.Y.; Urban population and the spread of medium-sized cities in Brazil. Interações, Campo Mourão, v. 14, n. 2, p. 251-265, jul./dez. 2013.

SOUZA, F.J.; BORSATO, A.V.; Bioclimatic characterisation of Campo Mourão. Geonorte Magazine, Special Issue 2, v.1, n.5, p.88 - 98, 2012.

VECCHI, B.P.T.; A study of mathematical models for the expansion of electricity distribution networks. 2004, p.144 Dissertation (Master of Science). Federal University of Paraná, Curitiba, 2004.

APPENDIX A-Delimitation of streets by intersections

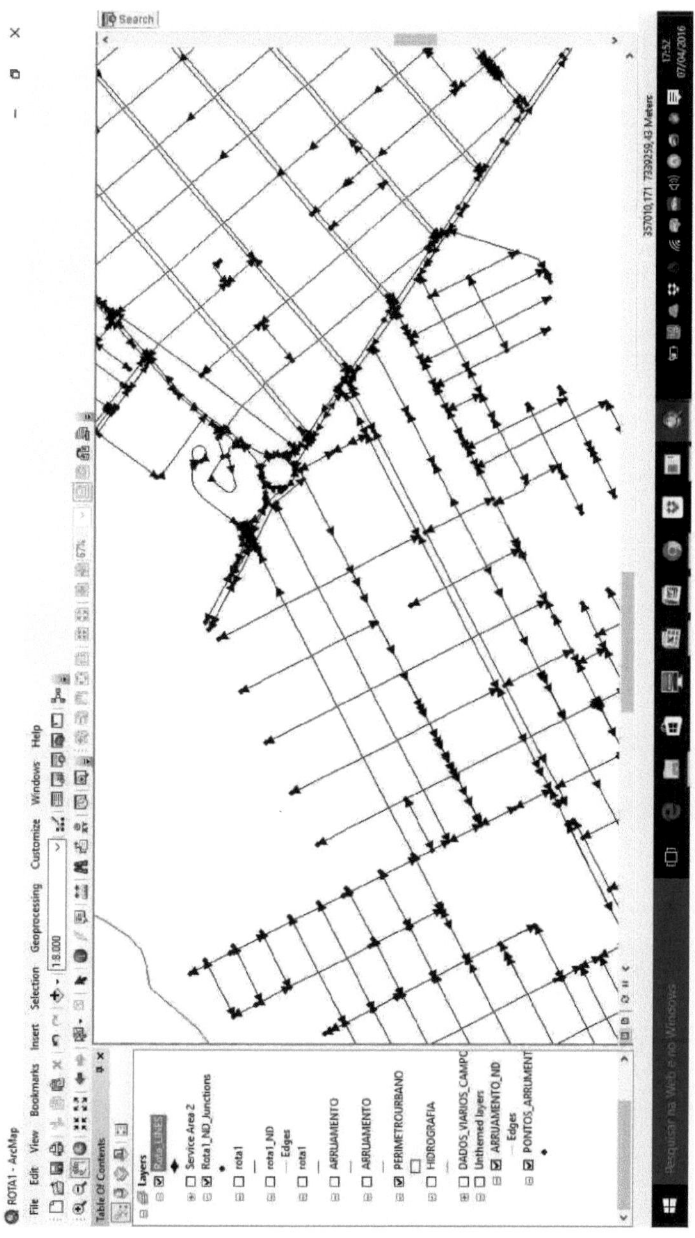

APPENDIX B - Allen Index spreadsheet data

Table

Rota_LINES

FID	Shape *	ObjectID	FacilityID	FromCumul	ToCumul_Le	SourceID	SourceOID	FromPositi	ToPosition	M	Allen	Expressão
0	Polyline M	9199	1	0	535,811186	1	1186	0	1	535,738 m	536	0
1	Polyline M	9200	1	535,811186	746,807335	1	575	0	1	210,967 m	211	0
2	Polyline M	9201	1	535,811186	1000	1	264	0	0,333564	464,130 m	464	0
3	Polyline M	9202	2	0	1000	1	54	0	0,988005	999,855 m	1000	0
4	Polyline M	9203	3	0	535,811186	1	1186	1	0	535,738 m	536	0
5	Polyline M	9204	3	0	210,996148	1	575	0	1	210,967 m	211	0
6	Polyline M	9205	3	0	1000	1	264	0	0,718595	999,880 m	1000	0
7	Polyline M	9206	4	0	1000	1	265	0	0,718339	1000,010 m	1000	0
8	Polyline M	9207	5	0	758,419304	1	54	0,250678	1	758,548 m	759	0
9	Polyline M	9208	5	0	253,721815	1	54	0,250678	0	253,452 m	253	0
10	Polyline M	9209	6	0	1000	1	265	0,142448	0,860787	1000,146 m	1000	0
11	Polyline M	9210	6	0	198,301767	1	265	0,142448	0	198,102 m	198	0
12	Polyline M	9211	7	0	1000	1	460	0	0,5527	999,406 m	999	0
13	Polyline M	9212	8	0	1000	1	460	0,008542	0,561243	999,082 m	999	0
14	Polyline M	9213	8	0	15,455451	1	460	0,008542	0	15,731 m	16	0
15	Polyline M	9214	9	0	921,681494	1	265	0,33792	1	921,598 m	922	0
16	Polyline M	9215	9	0	470,418425	1	265	0,33792	0	470,307 m	470	0
17	Polyline M	9216	10	0	270,54967	1	741	0	1	270,512 m	271	0
18	Polyline M	9217	11	0	1000	1	460	0,145713	0,696413	997,855 m	998	0
19	Polyline M	9218	11	0	263,637606	1	460	0,145713	0	264,072 m	264	0
20	Polyline M	9219	12	0	683,538314	1	264	0,491187	0	683,454 m	683	0
21	Polyline M	9220	12	0	708,06702	1	264	0,491187	1	707,956 m	708	0
22	Polyline M	9221	12	708,06702	932,724913	1	616	0	1	224,626 m	225	0
23	Polyline M	9222	12	683,538314	1000	1	1186	1	0,409378	316,419 m	316	0
24	Polyline M	9223	12	683,538314	894,534462	1	575	0	1	210,967 m	211	0
25	Polyline M	9224	12	932,724913	1000	1	1080	0	0,149829	67,067 m	67	0
26	Polyline M	9225	12	708,06702	1000	1	1108	1	0,396383	291,938 m	292	0
27	Polyline M	9226	12	932,724913	943,65163	1	44	0	1	10,925 m	11	0
28	Polyline M	9227	12	932,724913	956,214971	1	635	1	0	23,487 m	23	0
29	Polyline M	9228	12	708,06702	1000	1	1273	0	0,476751	291,901 m	292	0
30	Polyline M	9229	13	0	683,871339	1	265	0,508748	1	683,760 m	684	0
31	Polyline M	9230	13	0	708,228581	1	265	0,508748	0	708,144 m	708	0
32	Polyline M	9231	14	0	734,67222	1	264	0,527931	0	734,582 m	735	0
33	Polyline M	9232	14	0	656,933114	1	264	0,527931	1	656,828 m	657	0
34	Polyline M	9233	14	656,933114	881,591006	1	616	0	1	224,626 m	225	0
35	Polyline M	9234	14	734,67222	1000	1	1186	1	0,504811	265,292 m	265	0
36	Polyline M	9235	14	734,67222	945,668369	1	575	0	1	210,967 m	211	0
37	Polyline M	9236	14	881,591006	1000	1	1080	0	0,26371	118,191 m	118	0
38	Polyline M	9237	14	656,933114	1000	1	1108	1	0,290656	343,098 m	343	0
39	Polyline M	9238	14	881,591006	892,517724	1	44	0	1	10,925 m	11	0
40	Polyline M	9239	14	881,591006	905,081064	1	635	1	0	23,487 m	23	0
41	Polyline M	9240	14	656,933114	1000	1	1273	0	0,560257	343,029 m	343	0
42	Polyline M	9241	15	0	270,54967	1	741	1	0	270,512 m	271	0
43	Polyline M	9242	16	0	343,814578	1	891	0	1	343,766 m	344	0
44	Polyline M	9243	16	343,814578	385,714417	1	1033	0	1	41,894 m	42	0

0 (0 out of 9198 Selected)

Rota_LINES

APPENDIX C - Allen index spreadsheet data

Table

Rota_LINES

FID	Shape *	ObjectID	FacilityID	FromCumul	ToCumul_Le	SourceID	SourceOID	FromPositi	ToPosition	M	Allen	Expressão
0	Polyline M	9199	1	0	535,811186	1	1186	0	1	535,738 m	536	0
1	Polyline M	9200	1	535,811186	746,807335	1	575	0	1	210,967 m	211	0
2	Polyline M	9201	1	535,811186	1000	1	264	0	0,333564	464,130 m	464	0
3	Polyline M	9202	2	0	1000	1	54	0	0,988005	999,855 m	1000	0
4	Polyline M	9203	3	0	535,811186	1	1186	1	0	535,738 m	536	0
5	Polyline M	9204	3	0	210,996148	1	575	0	1	210,967 m	211	0
6	Polyline M	9205	3	0	1000	1	264	0	0,718595	999,880 m	1000	0
7	Polyline M	9206	4	0	1000	1	265	0	0,718339	1000,010 m	1000	0
8	Polyline M	9207	5	0	758,419304	1	54	0,250678	1	758,548 m	759	0
9	Polyline M	9208	5	0	253,721815	1	54	0,250678	0	253,452 m	253	0
10	Polyline M	9209	6	0	1000	1	265	0,142448	0,860787	1000,146 m	1000	0
11	Polyline M	9210	6	0	198,301767	1	265	0,142448	0	198,102 m	198	0
12	Polyline M	9211	7	0	1000	1	460	0	0,5527	999,406 m	999	0
13	Polyline M	9212	8	0	1000	1	460	0,008542	0,561243	999,082 m	999	0
14	Polyline M	9213	8	0	15,455451	1	460	0,008542	0	15,731 m	16	0
15	Polyline M	9214	9	0	921,681494	1	265	0,33792	1	921,598 m	922	0
16	Polyline M	9215	9	0	470,418425	1	265	0,33792	0	470,307 m	470	0
17	Polyline M	9216	10	0	270,54967	1	741	0	1	270,512 m	271	0
18	Polyline M	9217	11	0	1000	1	460	0,145713	0,698413	997,855 m	998	0
19	Polyline M	9218	11	0	263,637606	1	460	0,145713	0	264,072 m	264	0
20	Polyline M	9219	12	0	683,538314	1	264	0,491187	0	683,454 m	683	0
21	Polyline M	9220	12	0	708,06702	1	264	0,491187	1	707,956 m	708	0
22	Polyline M	9221	12	708,06702	932,724913	1	616	0	1	224,626 m	225	0
23	Polyline M	9222	12	683,538314	1000	1	1186	1	0,409378	316,419 m	316	0
24	Polyline M	9223	12	683,538314	894,534462	1	575	0	1	210,967 m	211	0
25	Polyline M	9224	12	932,724913	1000	1	1080	0	0,149829	67,067 m	67	0
26	Polyline M	9225	12	708,06702	1000	1	1108	1	0,396383	291,938 m	292	0
27	Polyline M	9226	12	932,724913	943,65163	1	44	0	1	10,925 m	11	0
28	Polyline M	9227	12	932,724913	956,214971	1	635	1	0	23,487 m	23	0
29	Polyline M	9228	12	708,06702	1000	1	1273	0	0,476751	291,901 m	292	0
30	Polyline M	9229	13	0	683,871339	1	265	0,508748	1	683,760 m	684	0
31	Polyline M	9230	13	0	708,228581	1	265	0,508748	0	708,144 m	708	0
32	Polyline M	9231	14	0	734,67222	1	264	0,527931	0	734,582 m	735	0
33	Polyline M	9232	14	0	656,933114	1	264	0,527931	1	656,828 m	657	0
34	Polyline M	9233	14	656,933114	881,591006	1	616	0	1	224,626 m	225	0
35	Polyline M	9234	14	734,67222	1000	1	1186	1	0,504811	265,292 m	265	0
36	Polyline M	9235	14	734,67222	945,668369	1	575	0	1	210,967 m	211	0
37	Polyline M	9236	14	881,591006	1000	1	1080	0	0,26371	118,191 m	118	0
38	Polyline M	9237	14	656,933114	1000	1	1108	1	0,290656	343,098 m	343	0
39	Polyline M	9238	14	881,591006	892,517724	1	44	0	1	10,925 m	11	0
40	Polyline M	9239	14	881,591006	905,081064	1	635	1	0	23,487 m	23	0
41	Polyline M	9240	14	656,933114	1000	1	1273	0	0,560257	343,029 m	343	0
42	Polyline M	9241	15	0	270,54967	1	741	1	0	270,512 m	271	0
43	Polyline M	9242	16	0	343,814578	1	891	0	1	343,766 m	344	0
44	Polyline M	9243	16	343,814578	385,714417	1	1033	0	1	41,894 m	42	0

0 (0 out of 9198 Selected)

Rota_LINES

Printed by Books on Demand GmbH, Norderstedt / Germany